Beginning PhoneGap

Mobile Web Framework for JavaScript and HTML5

Rohit Ghatol

Yogesh Patel

Apress®

Beginning PhoneGap: Mobile Web Framework for JavaScript and HTML5

ISBN-13 (pbk): 978-1-4302-3903-1

ISBN-13 (electronic): 978-1-4302-3904-8

President and Publisher: Paul Manning
Lead Editors: Michelle Lowman
Technical Reviewer: Dave Caolo
Editorial Board: Steve Anglin, Mark Beckner, Ewan Duckingham, Gary Cornell, Morgan Ertel, Jonathan Gennick, Jonathan Hassell, Robert Hutchinson, Michelle Lowman, James Markham, Matthew Moodie, Jeff Olson, Jeffrey Pepper, Douglas Pundick, Ben Renow-Clarke, Dominic Shakeshaft, Gwenan Spearing, Matt Wade, Tom Welsh
Coordinating Editor: Kelly Moritz
Copy Editors: Scribendi, Inc.
Compositor: MacPS, LLC
Indexer: SPi Global
Artist: SPi Global
Cover Designer: Anna Ishchenko

Distributed to the book trade worldwide by Springer Science+Business Media, LLC., 233 Spring Street, 6th Floor, New York, NY 10013. Phone 1-800-SPRINGER, fax (201) 348-4505, e-mail orders-ny@springer-sbm.com, or visit www.springeronline.com.

For information on translations, please e-mail rights@apress.com, or visit www.apress.com.

Apress and friends of ED books may be purchased in bulk for academic, corporate, or promotional use. eBook versions and licenses are also available for most titles. For more information, reference our Special Bulk Sales–eBook Licensing web page at www.apress.com/bulk-sales.

Any source code or other supplementary materials referenced by the author in this text is available to readers at www.apress.com. For detailed information about how to locate your book's source code, go to http://www.apress.com/source-code/.

This book is dedicated to the countless developers across the world who all worked hard to make HTML5/JavaScript/CSS the new standard for developing applications.

I would like to acknowledge the patience shown by my wife, Manija Ghatol, and the forgiving nature of my parents, S.G Ghatol and Manda Ghatol, toward whom I ignored my duties during the tenure of writing this book. I am fortunate to have a family that understands and supports my aspirations. Also, I would like to recognize the timely help provided by my friend and colleague Yogesh Patel, who is not only a technical genius but also a wonderful person.

— *Rohit Ghatol*

To my wife, Smita, who has always been there for me. Without you I am not in the place I am right now.

I wish to express my gratitude to Rohit for his immense support in my professional and personal life; you are one of the most inspiring people I've ever met.

—*Yogesh Patel*

Contents at a Glance

Contents

About the Authors

 **Rohit Ghatol** is a technology evangelist at heart. He loves to try new technologies and weave the vision of the future around them. He is currently working as an architect with QuickOffice and Synerzip. Rohit has more than 10 years of experience and is a regular technical speaker on various platforms in India including Indic Threads, SaltMarch, and CSI. He himself runs one such platform named TechNext on meetup. Rohit is an ex-Googler and worked with the OpenSocial team for a while.

 **Yogesh Patel** is a seasoned developer, building Web 2.0-based applications over the last 8 years. His expertise lies in Usability and Web 2.0 on JavaScript and GWT, and he is a corporate trainer on JavaScript and GWT-based technology. Yogesh is currently working as a Project Lead and Usability Guru with FuelQuest Inc. and Synerzip India, and he is actively involved in developing PhoneGap-based mobile applications with provision for Push from the cloud.

About the Technical Reviewers

Giacomo Balli is an iOS entrepreneur and freelance developer with successful apps on the App Store. He is an expert in HTML5/CSS3/AJAX modern techniques for webapps and an active reviewer of webkit with Apple. Giacomo loves working on new concepts and finding simple solutions to common problems. He has had a thorough exposure to PhoneGap multi-platform development since the first days.

Markus Leutwyler is a web and mobile developer living in Switzerland. He's interested in many web-based technologies and believes PhoneGap is one of the best solutions when creating universal mobile applications.

Nick McCloud started out with a Commodore PET and has programmed pretty much every desktop computer in a wide range of languages since. With experience pre-dating the world-wide-web, he has taken a keen interest in the application of web technologies to provide small enterprises with a competitive edge. Using PhoneGap, Nick has developed multiple cross-device apps for his clients.

Acknowledgments

Writing a book can be a tremendous responsibility, and when I began, the idea excited me as well as frightened me. Without the help and goodwill of various people, this book would not have come alive. The task was made easy by the Apress team. To name a few: Kelly Moritz really helped me understand the process and made things easier for me, and I got excellent and timely advice from Richard Carey and all the technical reviewers. Without the collaboration of these people, it would probably have become almost impossible. I have included the PhoneGap GWT Library from Daniel Kurka, and I would like to acknowledge his work and his help in understanding the library. Also, Yogesh Patel would like to acknowledge his student intern Omkar Ekbote, who helped develop and test the plug-in source code.

Finally, I would like to acknowledge the support from my colleague Nikhil Walvekar, a developer-cum-photography expert who helped me with some of the images in the book.

—Rohit Ghatol

Introduction

Who This Book Is For

This book is meant for anyone wanting to start mobile application development across more than one mobile platform. The book provides an introduction and detailed tutorial on PhoneGap and also helps the reader with the following:

1. Identifying which JavaScript UI Framework is best for them
2. Introduces the JavaScript UI Framework and its integration with PhoneGap
3. Explains the concept of a plug-in and how to use it to do OAuth authentication and Cloud Push
4. Explains how to write customized plug-ins

How This Book Is Structured

The book begins by explaining about the fragmentation in the mobile OS world and how it affects us. It goes further to talk about how to bridge the gap due to this fragmentation and how to write code once and deploy it across mobile platforms.

After the concept behind PhoneGap is made clear, the book goes on to explain PhoneGap usage on Android and then gives instructions on how to do the same across the other remaining mobile platforms.

It next introduces how to use a JavaScript UI Framework on top of PhoneGap and also talks about which JavaScript UI Framework to use in which scenario.

Finally, the book moves its focus to plug-ins. It shows a couple of examples of how to extend the PhoneGap framework with community plug-ins. Then it explains how to build these plug-ins across iOS, Android, and BlackBerry.

Downloading the Code

All the source code referred to in this book is available at https://bitbucket.org/rohitghatol/apress-phonegap. The chapters themselves state this. It is also available on the Apress web site at Apress.com.

Contacting the Authors

The authors can be contacted at their LinkedIn Profiles:
Rohit Ghatol—http://in.linkedin.com/in/rohitghatol
Yogesh Patel—www.linkedin.com/profile/view?id=19911394

Chapter 1

Understanding Cross-Platform Mobile Application Development

This book is about mobile application development; more specifically, about easing the pain of mobile application development. There are many smartphone platforms on the market: Android, iPhone, BlackBerry, Nokia, the Windows 7 Phone, and WebOS. Newer platforms are on the rise as well, such as Samsung's Bada and Meego.

The sheer number of development platforms for mobile applications may seem overwhelming. This is the first of many points you must keep in mind when dealing with mobile application development.

In the year 2000, we saw a similar situation in the desktop world. We had Microsoft Windows, Apple's Mac, and various versions of Linux and UNIX. At that time, it was difficult to build products that would run on all these platforms. The resulting fragmentation was often solved via in-house solutions by building frameworks in C++, with Operating System (OS)-specific modules abstracted. Fortunately, Sun's Java came to the rescue and provided us with a common platform on which to build. With Java's build–once–and–run–anywhere strategy, building desktop products had become a breeze.

Between 2004 and 2008, the developer community saw a different kind of fragmentation; this time, it took place in the browser world. It was a fragmentation involving the very popular Internet Explorer 6 vs. Firefox and Safari—then, Chrome and other browsers came out of the woodwork, causing further fragmentation.

The nature of this fragmentation, however, was different and a little more tame: it was mainly due to browsers not following the specifications outlined by the World Wide Web Consortium (W3C). Often, this fragmentation was solved by writing either "If Browser is IE, then do this else do that" or "If Feature is Present, then do this else do that."

Many JavaScript libraries came to the rescue and helped write cross-browser web applications. Things have improved to such an extent that all of the browsers are working hard to be more and more compliant with W3C specs. The browser, as a platform, is now a strong contender.

This book is about fragmentation in the mobile world. Mobile OS fragmentation is severe because there are no specifications or standards in this development area.

In 2007, Apple and Google launched their mobile platforms. In 2008, both companies launched mobile app stores to allow smartphone users to download mobile applications. The era of mobile applications had begun; since then, there has been no looking back. The number of smartphone users has grown exponentially.

Companies started focusing on delivering services and content on the new smartphone platform. Businesses realized they needed to shift their focus to smartphone users. Not only was there an increase in the number of users, but the frequency of smartphone usage increased as well.

Imagine your developers working around to the clock to release the same product on the iPhone, Android, BlackBerry, WebOS, and Symbia—and now, let's add Samsung Bada to that list! You can see the challenge here. The OS platforms, starting with their development environments, are so fragmented. For the iPhone, you will need Mac machines, and for BlackBerry, you will need Windows. This chapter will talk about these things in greater detail.

Now, for those of you who are new to mobile application development, we will start by focusing on what it's like to create a mobile application. We will answer questions like "How is a mobile application different than traditional web-based or desktop-based applications?" We will investigate the challenges of developing mobile applications for various platforms.

Types of Mobile Applications

It is important to understand the different types of mobile applications. I will put them in two categories, according to what they do.

1. Standalone mobile applications

2. Mobile applications (based on web services)

Standalone mobile applications are applications such as alarms, phone dialers, and offline games. Web service-backed mobile applications are applications like e-mails, calendars, Twitter clients, online games, and applications that interact with web services.

This distinction between mobile applications is unique to the context of this book. Although PhoneGap can be used to implement standalone mobile applications, the nature of PhoneGap-based mobile applications typically falls into the category of "service-backed mobile applications."

Understanding Web Services

As a developer, when you look at the web applications on the Internet, you need to think about two kinds of web development.

1. Web applications that are accessible via browsers (meant for human interfacing)

2. Web services that are accessible via protocols like RESTful web services (meant for programmatic interfacing)

All popular web applications like Google, Facebook, Twitter, LinkedIn, MySpace, Flickr, and Picasa provide a RESTful interface for their services. There are many online dictionaries for such sites. If you visit www.programmableweb.com, you will see a sizable listing of all of the web applications that provide such services for programmatic interfacing (see Figure 1–1).

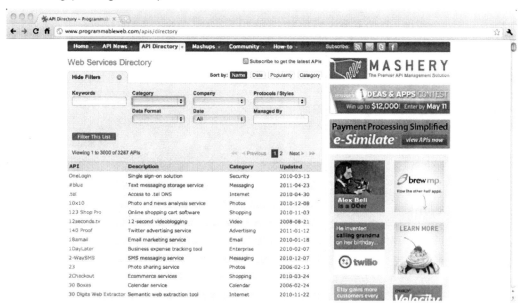

Figure 1–1. *Programmable Web API directory*

Many companies that want to develop mobile applications for multiple platforms either have their own web services or rely on other web services. While PhoneGap can work for standalone mobile applications, it is very well-suited for mobile applications that make use of web services. The reason for this is that PhoneGap applications are primarily web applications that are augmented with device features. Think about a Flickr web application that has access to a device's camera or Google Maps application, which, in turn, has access to a GPS. Another example is Foursquare, which has access to your GPS, as well as your phone's address book.

This more or less means that a majority of PhoneGap-based applications will access web services using JavaScript. This makes it important for developers using PhoneGap to have a handle on using web services.

For developers who want to write PhoneGap applications after reading this book, I recommend finding some web services on ProgrammableWeb.com, and writing a PhoneGap client for those services as an exercise.

This book will provide an example of one such service; namely, AlternativeTo.Net.

Overview of Mobile Applications

While many of you have at least some prior experience working with mobile applications, a large number of you are more familiar with non-mobile Platforms, (e.g., web platforms). Therefore, this book explicitly deals with the nature of mobile applications and the challenges associated with them. This will help you, if you come from a non-mobile background, in the way of understanding what it means to develop mobile applications.

Mobile Application Features

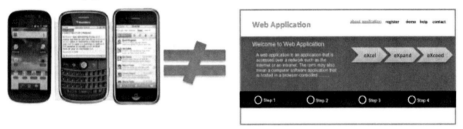

Figure 1–2. *Mobile applications are not web applications.*

The first thing to note is that mobile applications are not web applications. The difference is in both the nature of the features and the number of features provided (see Figure 1–3).

- A mobile application is likely to have fewer features.

- You can expect your mobile application to look very different from your web application. First, the screen size on your smartphone is not the same as your desktop. On a web application, where the screen is bigger, you have more space for menus, toolbars, and widgets.

  - Given the screen size constraint on your smartphone, you will see more of a dashboard type of home screen.

  - The smartphone user is expected to go through various levels of navigation to reach the feature he or she intends to use.

- Smartphone users and web users have different intentions. The smartphone user wants to use the application on the go, getting maximum productivity with the least amount effort, while the web user will likely spend more time using the web application.

Due to the preceding differences, you will see the most productive (or most frequently used) features being highlighted on smartphones. Whether a mobile application provides all the features, or a subset thereof, these small sets of productive (and most frequently used) features would be organized in the most accessible way on the mobile application.

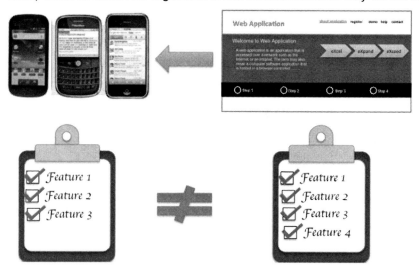

Figure 1–3. *Mobile features are not thesame as web application features.*

User Interaction

The way a user interacts with a mobile application relative to a traditional web application is very different (see Figure 1–4).

With the touch screen capabilities of a smartphone and more vivid user interaction, based on an accelerometer and compass, a mobile application has to be built differently.

Think about a car game application, where the car is maneuvered by tilting the phone to the left or right. This is based on an accelerometer. Think about a map application that always points north as the user changes his or her direction. This is based on a compass.

While the newer way to interact with applications has enhanced the user's experience, the absence of a physical keyboard on the newer mobile platforms adds some additional constraints for the power keyboard user. This needs to be taken into consideration when the mobile application requirements are being elaborated.

To add to this, a smartphone has two display modes: Layout and Portrait; these were unheard of in earlier browsers. An important part of documenting the requirement specification is to define the application's look, feel, and behavior when the device is in Portrait or Landscape mode.

Input Methods
- *Touch Screen*
- *DPad*
- *TrackBall*
- *Keyboard*
- *Accelerometer*
- *Compass*

Input Methods
- *Keyboard*
- *Mouse*
- *TrackPad*

Figure 1–4. *Smartphones and web applications have different User Input Interfaces.*

Location Awareness

Location awareness is something that comes naturally to a smartphone. Google Maps, Local Search, Foursquare, and many other mobile applications make use of the fine-grained GPS of smartphones. Web applications use location awareness too; however, these applications use relatively more course-grained GPS systems (e.g., country level) (see Figure 1–5).

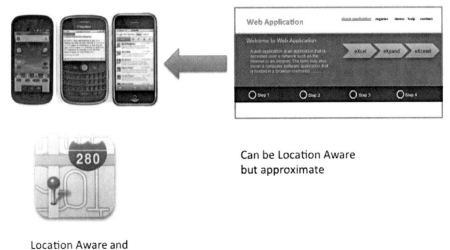

Can be Location Aware
but approximate

Location Aware and
highly accurate

Figure 1–5. *Location awareness capacity of smartphone apps compared to web applications*

Push Notification

Application users like to be notified of useful events like incoming e-mails and messages. A smartphone is the best platform for notification, since it's close to the user almost all of the time.

Apart from notifications like incoming e-mails or messages, any service can send notifications to a smartphone user (see Figure 1–6). Think about a workflow at an organization. Instead of a user always logging on to a web application to complete a workflow that involves him or her, it would be much more productive for the application to notify the user that he or she needs to perform an action to complete a workflow. This way, the user is always productive, irrespective of whether he or she is close to his or her laptop or desktop.

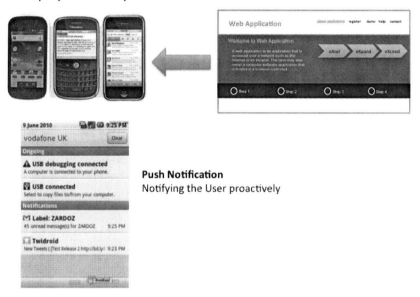

Push Notification
Notifying the User proactively

Figure 1–6. *Push notification capability of smartphones (notification on the go)*

Challenges in Cross-Platform Mobile Application Development

While mobile application development is exciting, given the growing number of mobile operating systems (OS), there are many challenges associated with developing mobile applications.

Let's take a look at those challenges.

OS Fragmentation

The trend of increased fragmentation coincides with the growing number of mobile platforms (see Figure 1–7). First, there were BlackBerry and Symbian smartphones— then came the powerful iPhone and Android platforms. To be sure, mobile platforms did not stop there. HP came with WebOS; Microsoft introduced the Windows 7 Phone; and now, Samsung is coming up with Bada.

This means that companies have to keep launching new products to make their presence felt on all mobile platforms.

Figure 1–7. *Fragmentation due to the growing number of mobile operating systems*

Let's say you want to develop a mobile application and target it for the iPhone, Android, BlackBerry, etc. Due to each mobile platform's different OS, consider the following:

- First, you have to set up different environments for each platform.

- Second, you need a bit of expertise with each respective OS. For a mobile developer, the learning curve may be long.

- Different programming languages are required for different mobile platforms.

- You need to be familiar with the features supported by each mobile platform; see Figure 1–10.

Table 1–1 depicts the required setup for mobile application development (for various mobile platforms).

In the past, we have seen similar OS fragmentations, beginning with the cross-desktop fragmentation of Windows, Linux, and Mac, which was resolved with Sun's launch of Java. In the more recent past, we faced browser fragmentation, which is resolved by cross-browser JavaScript frameworks like jquery, YUI, and Google Web Toolkit.

Mobile OS fragmentation is the worst and most diverse fragmentation of all. This adds a sizable technical challenge to launching mobile applications on all mobile platforms.

Multiple Teams/Products

If we choose to build a mobile application for each platform using multiple teams, we face a number of problems; adding teams leads to more risks with project delivery; adding products means more responsibilities for the product management team (see Figure 1–8). Since features are also fragmented on all mobile platforms, product management has to make specific requirements for products on each platform.

Ultimately, adding more teams, increasing coordination between multiple teams, and adding multiple products will lead to added overhead for the management and development teams.

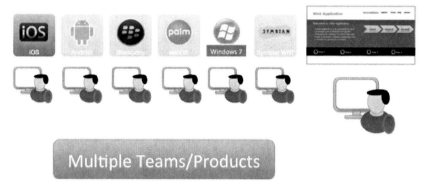

Figure 1–8. *Adding multiple teams for different mobile OSs poses new problems.*

Consistent User Experience

Given the fact that you want your application to be consistent across multiple mobile platforms, your application needs to give similar and consistent user experiences across all of the platforms (see Figure 1–9). This also has to do with the fact that your end-users could migrate from one platform to another, or maybe they are present on more than one platform. Think about a user who has an Android smartphone and an iPhone iPad. The user may use the iPad when he or she is at home or at the office, and may use the Android smartphone while he or she is on the go.

This is one of the many reasons why your application has to provide a similar user experience across mobile platforms; of course, user experience will vary to a degree depending on the mobile platform, due to the fragmentation of device features and capabilities.

Figure 1–9. *Providing a uniform user experience to application end-users across platforms*

Feature Fragmentation

Device features and capabilities vary across platforms (see Figure 1–10). This means that while some Androids and iPhones have an embedded compass to show directions, the other smartphones don't. This could mean that the navigation applications on other smartphones may not be able to rotate maps in the way that Android or iPhone applications can.

Overall, the fact that the same application will have some features turned off on some mobile platforms is a reality; the application's logic needs to be written in that manner.

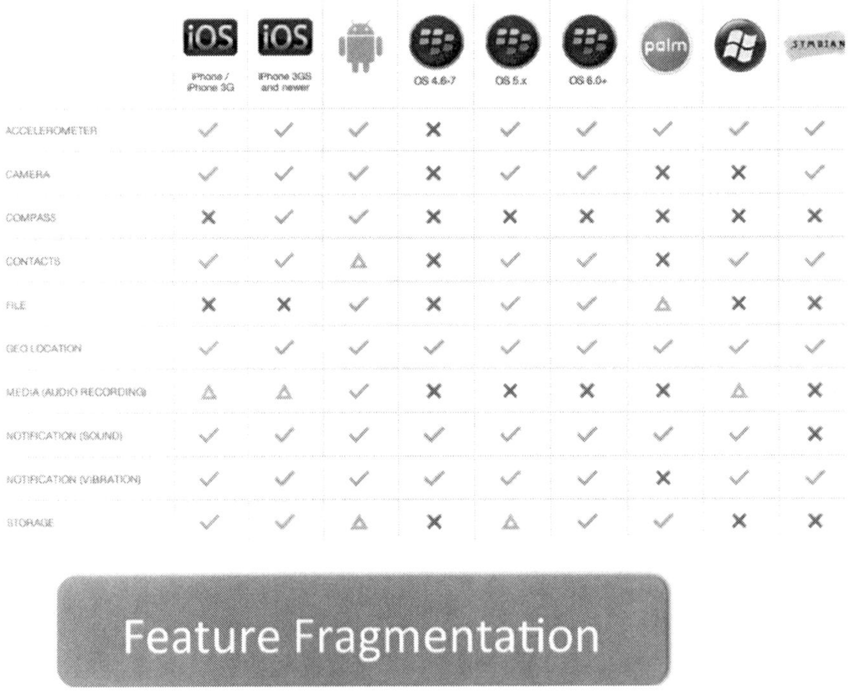

	iOS iPhone / iPhone 3G	iOS iPhone 3GS and newer	Android	BlackBerry OS 4.6-7	BlackBerry OS 5.x	BlackBerry OS 6.0+	palm	Windows	Symbian
ACCELEROMETER	✓	✓	✓	✗	✓	✓	✓	✓	✓
CAMERA	✓	✓	✓	✗	✓	✓	✗	✗	✓
COMPASS	✗	✓	✓	✗	✗	✗	✗	✗	✗
CONTACTS	✓	✓	△	✗	✓	✓	✗	✓	✓
FILE	✗	✗	✓	✗	✓	✓	△	✗	✗
GEO LOCATION	✓	✓	✓	✓	✓	✓	✓	✓	✓
MEDIA (AUDIO RECORDING)	△	△	✓	✗	✗	✗	✗	△	✗
NOTIFICATION (SOUND)	✓	✓	✓	✓	✓	✓	✓	✓	✗
NOTIFICATION (VIBRATION)	✓	✓	✓	✓	✓	✓	✗	✓	✓
STORAGE	✓	✓	△	✗	△	✓	✓	✗	✗

Figure 1–10. *Feature Fragmentation for different mobile OS's*

Development Environment Fragmentation

Development environment is one particularly important fragmentation. You will need at least two operating systems—Windows (preferably Windows 7) and Mac (preferably Leopard)—if you want to develop a mobile application targeting the following platforms:

1. iOS

2. Android

3. BlackBerry

4. WebOS

5. Symbian

6. Windows 7

What is more, you will have to use a variety of IDEs and programming languages, such as Java, C++, and Objective C. Also, you will be using a number of IDEs, such as Xcode and Eclipse.

Table 1–1 shows the requirements for development environments (for various mobile platforms).

Table 1–1. *Development Requirements*

Mobile OS	Operating System	Software/IDEs	Programming Language
iOS	Mac only	Xcode	Objective C
Android	Windows/Mac/Linux	Eclipse/Java/Android Development Tool (ADT)	Java
BlackBerry	Windows mainly	Eclipse/JDE, Java	Java
Symbian	Windows/Mac/Linux	Carbide.c++	C++
WebOS	Windows/Mac/Linux	Eclipse/WebOS plugin	HTML/JavaScript/C++
Windows 7 Phone	Windows mainly	Visual Studio 2010	C#, .NET, Silverlight or WPF

PhoneGap's Strategy for Cross-Platform Mobile Application

PhoneGap was made possible due to a commonality between all of the mobile platforms. If it were not for this common component, PhoneGap would not have been possible.

Browser Component As the Common Platform

The browser world was largely fragmented until just a few years ago. At the time, different browsers adhered to W3C standards to different degrees. Firefox and Safari browsers were at the forefront in terms of adhering to standards, while others lagged behind.

A lot has changed since then. Now, browsers are looking better in terms of adhering to standards (more so on the mobile platforms). This is also true because most modern mobile platforms have the same webkit-based browser.

Also, newer browsers, both on desktops and smartphones, have started to adhere to newer standards like HTML5/CSS3. This adds more features to the browser world and lessens the fragmentation across mobile platforms (see Figure 1–11).

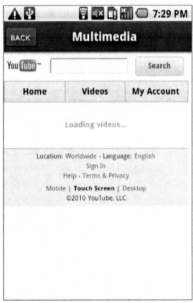

Figure 1–11. *Mobile browser*

Let's look at Table 1–2, which lists mobile platforms and their corresponding browser platforms. As you can see, all mobile platforms except the Windows 7 Phone use a webkit-based browser. While the Windows 7 Phone has its own browser, the good news is that all of the browsers listed here are already adhering to HTML5/CSS3 standards, and with the passage of time, their adherence will continue to improve.

Table 1–2 *Mobile Browsers*

Mobile OS	Browser
Android	Webkit-based
iPhone	Webkit-based
BlackBerry 6.0 +	Webkit-based
Windows 7 Phone	IE 7-based *
WebOS	Webkit-based
Nokia	Webkit-based
BADA	Webkit-based

PhoneGap uses these modern browsers as the platform for building HTML5/CSS3-based applications. Think of all PhoneGap applications as having embedded browsers and running these HTML5/CSS3-based applications.

Mobile Application Webviews

All of these mobile platforms support embedding browsers in applications. This means one of the screens of your mobile application can actually be a browser that shows an HTML page.

These embedded browsers are often referred as *webviews*. This means you can define one of the screens of your application as a webview.

Think about your application having a screen named "about us." This "about us" screen shows your company's information. Now, let's assume for example, the "about us" information about your company changes on a frequent basis. One of the requirements of your mobile application is to show the latest "about us" information. Therefore, instead of showing a hardcoded "about us" screen, you can show a webview pointing to your company's "about us" page (preferably the mobile version of the web page). It will load the "about us" page from the web. Also, a webview can be used to load and display the HTML pages that are stored locally on the mobile device. We can take this concept a step further: instead of a static web page, we can show Ajax-based web pages that interact with web services.

Native Hooks to Expose Device Capabilities

Now that we know that browsers can be embedded within a web application, let's shift our focus to exposing device capabilities through these embedded browsers.

Let's say you are developing a Flickr application, based on a Flickr API. With the help of these APIs, you can login to Flickr, list galleries, and download and show your pictures.

While this is a good idea for a web application, when we show the same application on a mobile phone, remember that a mobile phone usually has a camera. It would make perfect sense to allow the Flickr application to take a picture from the camera and upload it to Flickr.

In order to do this, we can make the embedded browser (or webview) expose JavaScript API, which, when called, makes the camera take a picture and gives us back the binary data for that picture (see Figure 1–12).

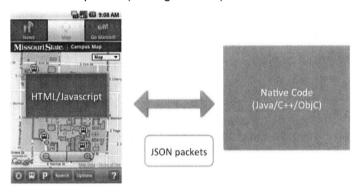

Figure 1–12. *JavaScript to native communication and vice versa*

Technically, all these platforms support exposing native modules to JavaScript in the webview. This means, programmatically, that all these platforms allow JavaScript code to call native Java/C++/Objective C code, and vice versa.

Let's take a look at an example. Our webview hosts an HTML page, which is showing a Google map. We want to center the map according to the GPS location of the phone. In order to do so, we need to write a native component, which enquires the device about the GPS location.

Then, we write code that will expose this native module from the webview. The JavaScript code in the webview invokes this code to gain access to the GPS coordinates. Once the code gains access to the GPS coordinates, it centers the map accordingly. This is the main principle behind the PhoneGap framework.

HTML5 and CSS3: The Standards for Writing Applications

HTML5 and CSS3 are emerging web technologies. They are making web applications more interactive and feature-rich.

HTML5 has not only added new markups for more robust multimedia support; it has also added features like web worker for background processing, offline support, database support, and much more.

CSS3 is the new standard for a seamless, rich User Interface (UI). Gone are the days when designers were put to task to get simple rounded corners or gradients on a button or border. With CSS3, things are easier, faster, and better.

With the support for animation, a CSS3 site can now compete against flash-based sites. Not only that, but a portal site can be easily transformed into a mobile site by a mere change of the CSS file. Furthermore, print previews can now be achieved with a different CSS file.

It's a well-known fact that mobile browsers are early adopters of W3C standards. This means mobile phones are the right platform for HTML5/CSS3 applications.

Single Origin Policy Not Applicable

For those of you who have worked with Ajax-based applications, you know that a web application hosted at "abc.com" cannot make Ajax calls to a web service hosted at "xyz.com." This means that if someone was developing an Ajax-based application—say, hosted at myphotobook.com—he or she would not be able to make Ajax calls to flickr.com.

This is called a *single origin policy*—you can read further about single origin policies at `http://en.wikipedia.org/wiki/Same_origin_policy`.

The same is not true for a PhoneGap application. A PhoneGap application bundles the required HTML, JavaScript, and CSS files, and PhoneGap applications do not have domains like "abc.com." This allows PhoneGap to be a platform for the easy development of mashups, which can freely make Ajax calls to various other sites.

Think about your PhoneGap application integrating Facebook, Twitter, and Flickr all into one mashup, with just a few lines of JavaScript code.

This makes PhoneGap an ideal platform for creating mobile applications for the web services listed on programmableweb.com.

The restrictions are illustrated in Figure 1–13:

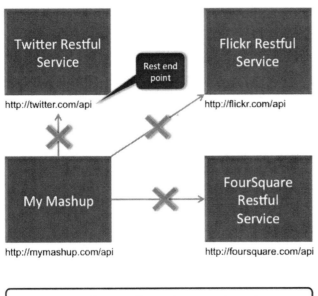

Figure 1–13. *Single origin policy*

Conclusion

PhoneGap uses HTML5, JavaScript, and CSS3 to develop mobile applications. These are standard technologies in the web world. By using PhoneGap, a developer with little or no native language background can start developing mobile applications for all of the popular mobile platforms.

Although PhoneGap provides access to standard native features of mobile applications, its plug-in framework is flexible enough to extend and add new features, if required.

PhoneGap is a growing technology used to develop cross-mobile platform applications.

Chapter 2

Getting Started with PhoneGap

PhoneGap is a HTML5 application framework that is used to develop native applications through web technologies. This means that developers can develop Smartphone and Tablet applications with their existing knowledge of HTML, CSS, and JavaScript. With PhoneGap, developers don't have to learn languages like Objective-C for the iPhone.

Applications that are developed using PhoneGap are hybrid applications. These applications are not purely HTML/JavaScript based, nor are they native. Parts of the application, mainly the UI, the application logic, and communication with a server, is based on HTML/JavaScript. The other part of the application that communicates and controls the device (phone or tablet) is based on the native language for that platform. PhoneGap provides a bridge from the JavaScript world to the native world of the platform, which allows the JavaScript API to access and control the device (phone or tablet).

PhoneGap essentially provides the JavaScript API with access to the device (phone or tablet) capabilities like, the camera, GPS, device information, and many others. These APIs are covered in detail in Chapter 4.

This chapter starts with providing you with proper information to understand the overall architecture of PhoneGap. Then we will apply this information in a PhoneGap example. At the end of this chapter we will write a small Hello World Application using PhoneGap.

> **NOTE:** PhoneGap is a framework; it does not provide any IDEs or special development environments for coding. You will need to use Eclipse and Android SDK to develop a PhoneGap application for an Android; you will need to use Xcode to develop a PhoneGap application for an iPhone.

PhoneGap Architecture

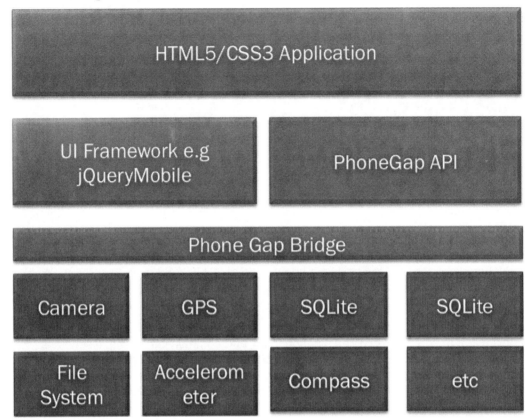

Figure 2–1. *PhoneGap application architecture*

The PhoneGap framework is primarily a JavaScript Library that allows HTML/JavaScript applications to access device features. The PhoneGap framework also has a native component, which works behind the scene and does the actual work on the device (phone or tablet).

Please refer to Figure 2–1 for overall PhoneGap architecture. An application build using PhoneGap will primarily have two parts:

1. The JavaScript Business Logic Part, which drives the UI and its functionality.

2. The JavaScript Part, which accesses and controls the device (phone or tablet).

Consider a Facebook application. The main parts of the application would be the login page, and downloading photo galleries. Now you want to add a module where you can take a picture and upload it to Facebook. In order to do this, you would call PhoneGap's camera API to gain access to the phone's camera, take a picture, and get the picture file. The next step is an AJAX call to the Facebook Server, in order to upload the picture.

Another example that can be applied is using PhoneGap to store a Friend List in a database, so we can search for local friends.

The previous description gives the impression that developing mobile applications in the PhoneGap requires more of writing business logic and UI, and less accessing of the device's capabilities, which is correct. This book not only explains PhoneGap APIs but also acts as a guide for creating a HTML5/CSS3 based mobile application.

Setting up an Environment on the Android

The first step towards creating a PhoneGap application is to setup a mobile development environment. We will begin with Android because the Android application development is in Java, which is based on Eclipse, and supports almost all features of PhoneGap.

You will need to download and install the following prerequisites for Android:

1. JDK 1.6+

2. Eclipse 3.4 to 3.6

3. Android SDK with an Android 2.2 platform

4. Android ADT plugin for Eclipse

5. Android AVD for Android 2.2

6. PhoneGap SDK 1.1.0 for Android

Since Android is programmed in Java, we need JDK 1.6+ and Eclipse 3.4+. We will then install Android SDK. The Android SDK is a generic SDK and does not come with support for any platform. A platform is an OS version, for example 2.2 Froyo, 2.3 Ginger Bread, and 3.0 Honeycomb. These platforms need to be downloaded in order to create, build, and run Android projects. This plugin is called the Android ADT Plugin.

Once the Eclipse, Android SDK, and Android ADT (Eclipse Plugin) are all set, we need to create an Emulator Environment for Android. This is called a Preparing Android AVD (Android Virtual Device). If we are developing a PhoneGap Application for Android that is targeting 2.2 Froyo, we need an AVD of the same Android platform.

The following steps will explain how to create an Android Project and inject the PhoneGap Library into the Android.

Required Installations for PhoneGap Android Project

1. Install the 3.4 version of Eclipse.

2. Install Android SDK.

3. Install the Android ADT Plugin for Eclipse.

4. Create AVD for the Emulator.

5. Install the PhoneGap libraries.

Step 1: Set-up Eclipse

This step assumes you already have Java SDK 1.6 installed. Once that has been installed, download Eclipse from www.eclipse.org/downloads/. See Figure 2–2 to see the eclipse download page. We need to have an Eclipse IDE version 3.4+ with support for JDT (Java Development Environment). You should install Eclipse IDE for Java Developers.

Figure 2–2. *Eclipse download page*

Step 2: Install Android SDK

Some of the steps in setting up the Android Development Environment are platform dependent. To avoid any confusion, we will explain how to execute each step in a platform specific manner.

Start by downloading Android SDK from http://developer.android.com/sdk/index.html (refer to Figure 2–3).

Figure 2–3. *Android SDK download page*

Instruction for Windows

Install Android SDK by using the Android Installer, installer r11-windows.exe. This is the recommended installation technique for Windows. The alternative is to download the android-sdk r11-windows.zip file, and extract it to a folder. We assume that the Android SDK is extracted to c:\android_sdk.

Instructions for Linux

Download the archie android-dk_r11-linux_x86.tgz archive and extract it to a folder.

Instruction for Mac OSX Intel

Download the archive android-sdk_r11-mac_x86.zip file and extract it to folder.

This Android SDK can support all Android platforms that have been released so far. These platforms include the Android 1.1 platform to the recent Android 3.0 (Honeycomb) platform. Since nobody requires all of the platforms, the Android SDK comes with no platform preinstalled.

For this book, we will focus only on the SDK platforms: Android 2.2, API 8, and revision 3.

Since there are no platforms preinstalled, the next step is to install the platforms you are interested in. Go to the Android SDK location (in our case c:\android_sdk), and open an executable named Android in the tools folder. In case you have bandwidth limitations, instead of downloading all of the platforms, download only the 2.2 Platform of Android (SDK platforms Android 2.2, API 8, and revision 3).

This will open the following screen seen in Figure 2–4. Select the Available Package option, check that Android Repository, and click Install.

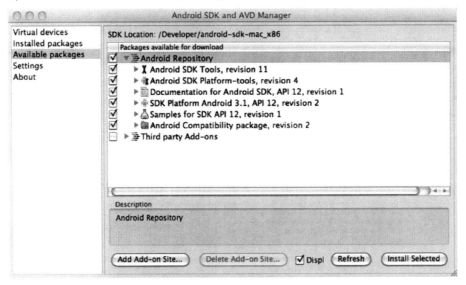

Figure 2–4. *Available platform packages that can be installed*

Now that you have downloaded the platforms, you have the necessary tools to create applications for all Android versions that have been launched so far.

It is recommended you install all of the available packages so that you can have the tools to create Android Projects for any of the Android platforms that have been released.

If you want to develop a mobile application for Froyo (Android 2.2), you need to have Froyo (Android 2.2.) listed in the installed packages.

Step 3: Install the Android ADT Plugin for Eclipse

1. Launch Eclipse and click on Help->Install New Software to open the Available Software Dialog box.

2. In the Work With text box, enter the URL `https://dl-ssl.google.com/android/eclipse`), as seen in Figure 2–5.

3. When you see the option to install Developer Tools, click on it, select all of the check boxes in the Developer Tools check boxes, and click on Next.

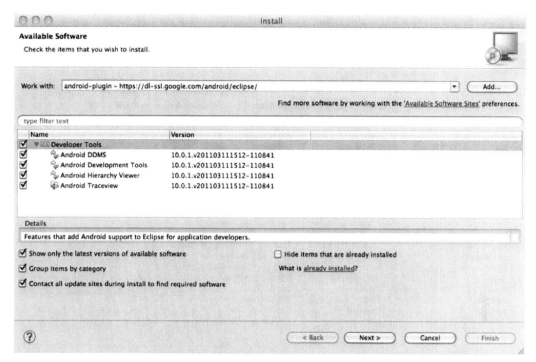

Figure 2–5. *Installing the Android ADT Plugin for Eclipse*

1. Configure the Android ADT Plugin with the location of the previously installed Android SDK. Open Eclipse's Preferences by clicking Windows->Preferences for Windows and Eclipse->Preferences for Mac. In case you receive an Unsigned Content Warning Dialog, you can safely ignore that.

2. In the Preferences pane, click and expand the Android option. You will see Android Preferences pane as shown in Figure 2–6. In the Android Preferences pane, put in the location of the Android SDK in the SDK Location text box, and hit Apply.

    If the Android SDK Location is correct, you should see a number of options under Target Name, including Android 2.2.

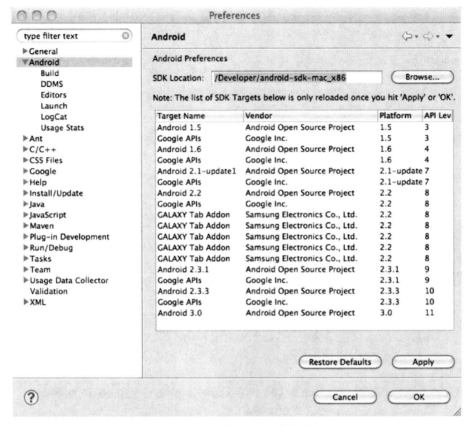

Figure 2–6. *Setting the Android SDK's location in the Android Preferences screen.*

Step 3: Create Android AVD for the Android 2.2 Platform

1. Open Eclipse and create a workspace for the Android PhoneGap. The next step is to create an emulator for Android. Since Android comes with many platform versions, we have to create an Android Virtual Device (AVD) for each platform that is targeted. In Figure 2–7, you will see your eclipse as depicted in the screen.

    Please note that the Android emulator runs an Android Virtual Device (AVD).

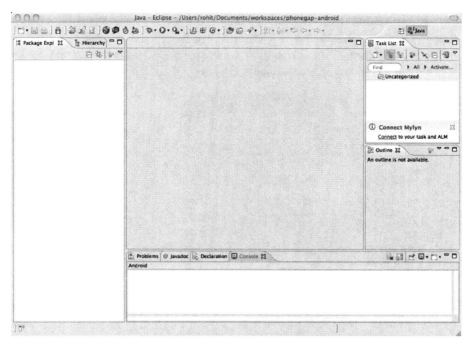

Figure 2–7. *Eclipse with the ADT Plugin.*

2. Click on the  button on the toolbar to open the Android SDK and the AVD Manager. Choose the Virtual Devices option as depicted in Figure 2–8.

Figure 2–8. *The Android SDK and AVD Manager*

3. Click on the New Button to create a new AVD. Choose the Android 2.2 platform, also known as Froyo. Choose the 128 MB SD Card Size, and choose the Skin Built-in as HVGA. After all of that has been filled out, click Create AVD. Refer to Figure 2–9 to see what the "AVD Screen" looks like.

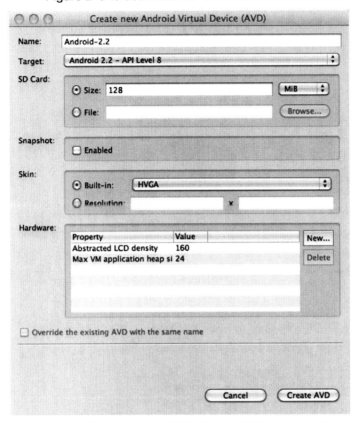

Figure 2–9. *Creating a new Android Virtual Device (AVD) to be run in the Android Emulator*

You will see the AVD that was created, depicted in Figure 2–10.

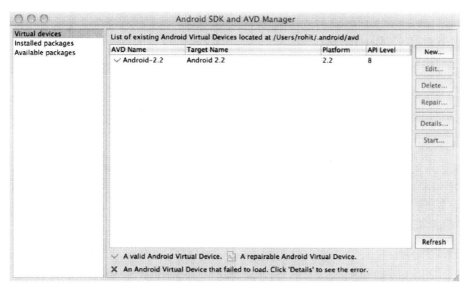

Figure 2–10. *AVD for the Android 2.2 Platform (Froyo)*

Step 4: Install the PhoneGap SDK

1. Download the PhoneGap SDK 1.1.0 from the following link,
   `http://phonegap.googlecode.com/files/phonegap-1.1.0.zip`. After this zip is
   extracted you should see a directory structure, as seen in Figure 2–11.

Figure 2–11. *PhoneGap SDK 1.1.0 directory structure*

2. Select the Android directory and you will see the phonegap-1.1.0.jar and the
   phonegap-1.1.0.js files (see Figure 2–12).

Figure 2–12. *Android folder within the PhoneGap SDK.*

This completes the setup of the PhoneGap for Android.

Create a New Project

The first application in this book is a Hello World Application. The Hello World PhoneGap mobile application shows a Hello World on the screen once the PhoneGap framework is loaded.

Step 1: Create an Android Project

Open Eclipse, click on File->New Project->Android Project. This will open up an Android Project dialog box as shown in Figure 2–13 and Figure 2–14. This is shown in the following steps:

1. Put PhoneGap-helloworld as the project name.

2. Ensure that you have selected Android 2.2 as the build target.

3. Enter Helloworld as the application name. This is the human readable name of the application.

4. Enter `org.examples.phonegap.sample` as the package name. An application in the Android market is uniquely identified by the package name. There cannot be two Android applications with the same package name on the Android market.

5. Check the Create Activity checkbox and enter helloworld as the activity name. The activity in Android is a screen. And the activity name is also the class name of the activity.

6. Put 7 in the min SDK version. This means that you will allow this application to be searched and installed by all Android 2.1 device platforms, also known as the Éclair Android phones.

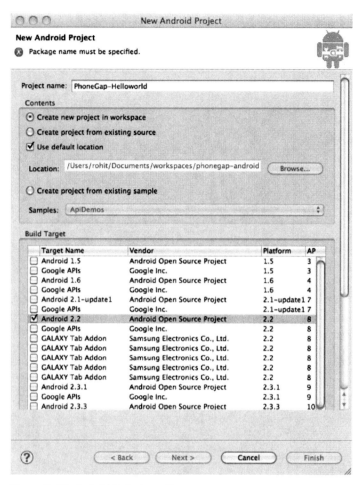

Figure 2–13. *Android Project creation*

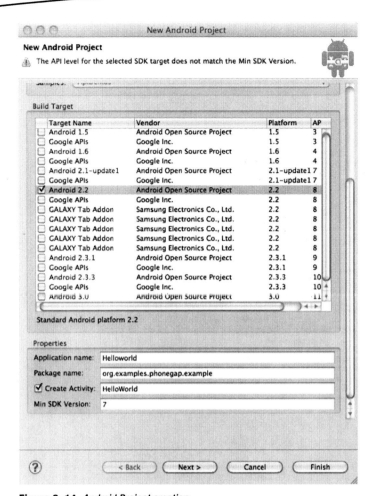

Figure 2–14. *Android Project creation.*

Step 2: Add PhoneGap Libraries to the Project

Once the Android Project is created, it's time to inject the PhoneGap framework into the Android Project. As we have mentioned before, PhoneGap comes with three main components: the native component, the XML plugin, and a JavaScript file.

1. To install the native component in Android, create a directory named lib in the project and copy the PhoneGap jar into it. You can either drag and drop the `phonegap-1.1.0.jar` in the lib folder, or you can copy and paste it into the lib folder in the Eclipse IDE. Next, add the PhoneGap jar to the class path by right clicking Build Path -> Add to Build Path. This is highlighted in Figure 2–15.

2. Copy the XML directory from the PhoneGap's Android Directory into the res folder.

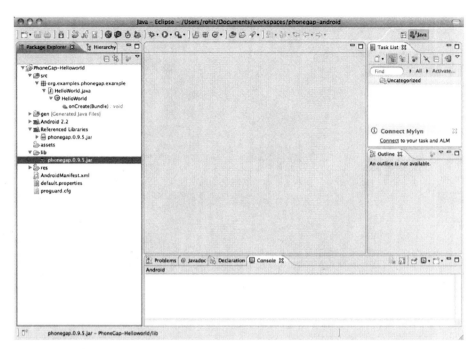

Figure 2–15. *Highlighting the location of the PhoneGap jar in the Android Project*

3. Once the PhoneGap Jar Is added to the Android Project, it's time to inject the JavaScript file of the PhoneGap into the project. We will create a www folder under the Assets Folder of the Android Project. The Assets Folder is like the media folder of the Android Application. In our case, we will put all of the files of the browser-based application inside of the www folder. To begin with, add the PhoneGap JavaScript file to the www folder, found in the Assets Folder. This is highlighted in Figure 2–16.

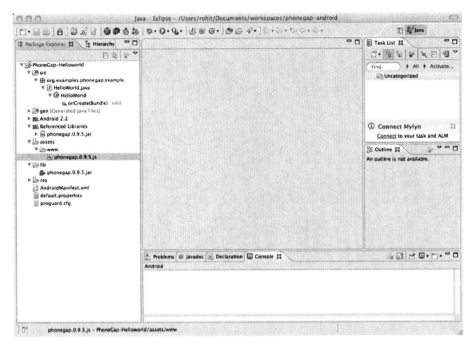

Figure 2–16. *Highlighting the location of the PhoneGap JavaScript file in the Android Project*

Step 3: Modify Android Permissions

In Android applications, the main file is the Android Manifest file. In this file there are many specific things, like the package name, which uniquely identify the application on the market. The main file contains a section called permissions. Android uses this section to inform the user that the application will be using certain features of the phone. Say an application intends to use the Internet to fetch data; permission needs to be attained in order to install the application. When the user installs the application he will be shown by the Android market that this application will be given the permission to use the Internet.

For the PhoneGap, the following permissions need to be added:

1. Add the following permissions to the Android Manifest XML:

```
<uses-permission android:name="android.permission.CAMERA" />
        <uses-permission android:name="android.permission.VIBRATE" />
        <uses-permission android:name="android.permission.ACCESS_COARSE_LOCATION" />
        <uses-permission android:name="android.permission.ACCESS_FINE_LOCATION" />
        <uses-permission
android:name="android.permission.ACCESS_LOCATION_EXTRA_COMMANDS" />
        <uses-permission android:name="android.permission.READ_PHONE_STATE" />
        <uses-permission android:name="android.permission.INTERNET" />
        <uses-permission android:name="android.permission.RECEIVE_SMS" />
        <uses-permission android:name="android.permission.RECORD_AUDIO" />
        <uses-permission android:name="android.permission.MODIFY_AUDIO_SETTINGS" />
        <uses-permission android:name="android.permission.READ_CONTACTS" />
```

```
<uses-permission android:name="android.permission.WRITE_CONTACTS" />
<uses-permission android:name="android.permission.WRITE_EXTERNAL_STORAGE" />
<uses-permission android:name="android.permission.ACCESS_NETWORK_STATE" />
```

2. We will also need to add the supports-screen option in the Manifest file, as seen as follows:

```
<supports-screens
        android:largeScreens="true"
        android:normalScreens="true"
        android:smallScreens="true"
        android:resizeable="true"
    android:anyDensity="true" />
```

3. Add android:configChanges=orignetation|keyboardHidden to the activity in the Android Manifest. This tells the Android not to kill and recreate the activity when the user flips the phone and the screen switches from portrait to landscape and vice versa.

4. Add a second activity after the previous one, by applying the following XML snippet:

```
<activity android:name="com.phonegap.DroidGap" android:label="@string/app_name"
android:configChanges="orientation|keyboardHidden">
        <intent-filter> </intent-filter>
</activity>
```

Once you have modified the Android Manifest, as per the previous instructions, an Android Manifest XML will appear. It will be seen as follows:

```
<?xml version="1.0" encoding="utf-8"?>
<manifest xmlns:android="http://schemas.android.com/apk/res/android"
    package="org.examples.phonegap.helloworld" android:versionCode="1"
    android:versionName="1.0">
    <supports-screens android:largeScreens="true"
        android:normalScreens="true" android:smallScreens="true"
        android:resizeable="true" android:anyDensity="true" />
    <uses-permission android:name="android.permission.CAMERA" />
    <uses-permission android:name="android.permission.VIBRATE" />
    <uses-permission android:name="android.permission.ACCESS_COARSE_LOCATION" />
    <uses-permission android:name="android.permission.ACCESS_FINE_LOCATION" />
    <uses-permission android:name="android.permission.ACCESS_LOCATION_EXTRA_COMMANDS"
/>
    <uses-permission android:name="android.permission.READ_PHONE_STATE" />
    <uses-permission android:name="android.permission.INTERNET" />
    <uses-permission android:name="android.permission.RECEIVE_SMS" />
    <uses-permission android:name="android.permission.RECORD_AUDIO" />
    <uses-permission android:name="android.permission.MODIFY_AUDIO_SETTINGS" />
    <uses-permission android:name="android.permission.READ_CONTACTS" />
    <uses-permission android:name="android.permission.WRITE_CONTACTS" />
    <uses-permission android:name="android.permission.WRITE_EXTERNAL_STORAGE" />
    <uses-permission android:name="android.permission.ACCESS_NETWORK_STATE" />
    <uses-sdk android:minSdkVersion="7" />

    <application android:icon="@drawable/icon" android:label="@string/app_name">
        <activity android:name="HelloWorld" android:label="@string/app_name"
            android:configChanges="orientation|keyboardHidden">
```

```
                    <intent-filter>
                        <action android:name="android.intent.action.MAIN" />
                        <category android:name="android.intent.category.LAUNCHER" />
                    </intent-filter>
            </activity>
            <activity android:name="com.phonegap.DroidGap"
android:label="@string/app_name"
                    android:configChanges="orientation|keyboardHidden">
                    <intent-filter>
                    </intent-filter>
            </activity>
        </application>
</manifest>
```

Step 4: Modify the Main Activity

In Android, a class named activity represents a screen. In order for us to use the
PhoneGap in the Android, we will change the screen from an activity to a DroidGap.
DroidGap is a special activity, which allows us to show HTML pages. This class is
shown In Figure 2–17 for the HelloWorld Class.

> **NOTE:** We are telling the DroidGap to load the index.html file in the Android Assets.

```
package org.examples.phonegap.helloworld;

import android.os.Bundle;

import com.phonegap.DroidGap;

public class HelloWorld extends DroidGap {
    /** Called when the activity is first created. */
    @Override
    public void onCreate(Bundle savedInstanceState) {
        super.onCreate(savedInstanceState);
        super.loadUrl("file:///android_asset/www/index.html");
    }
}
```

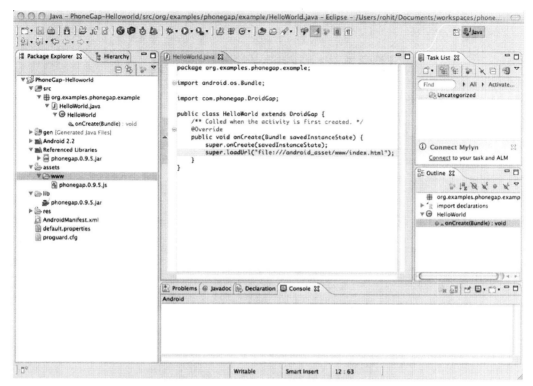

Figure 2–17. *Activity extending the DroidGap class*

Write the HelloWorld Application

A PhoneGap application is an HTML/JavaScript application. Refer to Figure 2–18. Following is the index.html.

1. Include the PhoneGap JavaScript Library version 1.1.0 in the HTML page.

2. Register the `init()` method with the body's onload event.

3. In the `init()` function, register the JavaScript callback function onDeviceReady with the deviceready event.

4. In the onDeviceReady callback function, change the contents of the h1 element with the ID "helloworld" with the text "hello World! Loaded PhoneGap Framework!"

The complete source code is listed here:

```
<!DOCTYPE HTML>
<html>
  <head>

    <title>PhoneGap</title>

    <script type="text/javascript" src="phonegap-1.1.0.js"></script>
```

```
<script type="text/javascript">

   /** Called when phonegap javascript is loaded */
   function onDeviceReady(){
     document.getElementById("helloworld").innerHTML
    ="Hello World! Loaded PhoneGap Framework!";
   }

   /** Called when browser load this page*/
   function init(){
       document.addEventListener("deviceready", onDeviceReady, false);
   }

</script>
</head>
<body onLoad="init()">

   <h1 id="helloworld">...</h1>

</body>
</html>
```

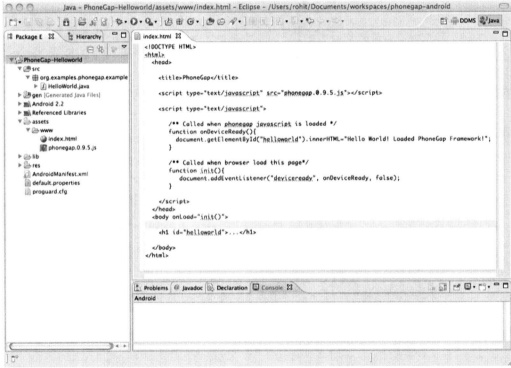

Figure 2–18. *Index.html of the PhoneGap Project*

You can download the complete source for this chapter from https://bitbucket.org/rohitghatol/apress-phonegap/src/67848b004644/android/PhoneGap-Helloworld

Deploy to Simulator

In order to run the Android application, right click on the project PhoneGap-helloworld, select Run As, and select Android Application.

This will launch the emulator with the AVD that we previously created. And you will see following screens as the application loads. As the application launches, you will see … on the screen, as depicted in Figure 2–19.

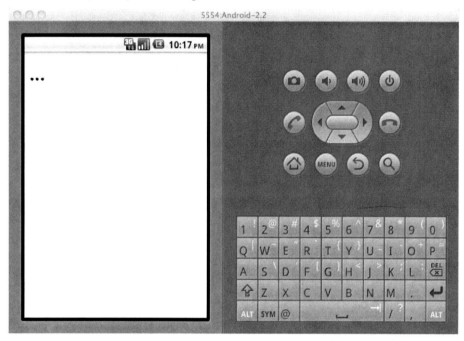

Figure 2–19. *The PhoneGap application loads with … for a few seconds.*

Once the PhoneGap framework is loaded, you will see the application display a message, as depicted in Figure 2–20.

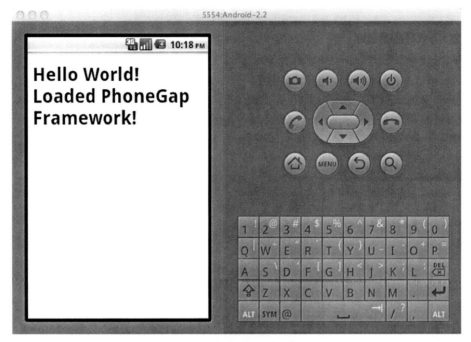

Figure 2–20. *The PhoneGap app shows a message after the PhoneGap framework loaded up the Deploy to Device application.*

So far we have seen how to test applications on the emulator. However, there are certain features that cannot be tested on the emulator. In order to test the GPS, camera, accelerometer, compass, and actual user perception, the actual device needs to be tested.

Deploy to the Device

Deploying an Android application to a device is a two-step process:

Step 1: Get the device ready.

1. Unlock your device and press the Menu key. This will give you a view that looks like Figure 2–21.

Figure 2–21. *Go to the Android phone's Settings*

2. Click on the Settings and the screen in Figure 2–22 will appear. Choose the Applications option.

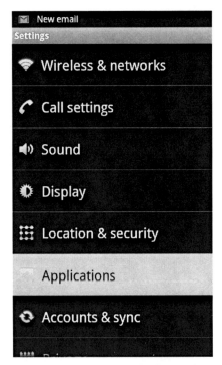

Figure 2–22. *Go to the Applications setting*

3. Now we must ensure that we can deploy non-Market applications on our device. This is done by clicking the Unknown sources, seen in Figure 2–23.

Figure 2–23. *Click the Unknown sources so we can add the non-Market applications*

4. The next step is to go into the Development options (see Figure 2–24) and enable the USB debugging (see Figure 2–25). This allows you to plugin one end of the USB cable to your Android device and the other end of your cable to your PC or Mac. Use Eclipse to debug the application running on your device.

Figure 2–24. *Go into the Development options*

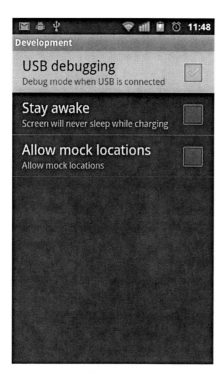

Figure 2–25. *Enable the USB debugging*

Now your device is all set to deploy applications.

The Android ADT Plugin provides the Android with a Dalvik Debug Monitoring Server (DDMS). DDMS has many features, such as listing the devices/emulators that are currently available to deploy and debug an Android application, allowing users to see the log messages from the application that was deployed on the device/emulator, and browsing the file system of the device/emulator.

Step 2: At the Application Launch Type, provide information that we intend to deploy to the device.

1. Plug the USB cable into your device, and plug the USB end into your development machines. Now open Eclipse and go into the DDMS perspective (Go to Eclipse->Windows->Open Perspective->DDMS). You will see a screen like Figure 2–26. This screen depicts that we have an Android emulator running, and that we also have an Android device plugged into the machine's USB input.

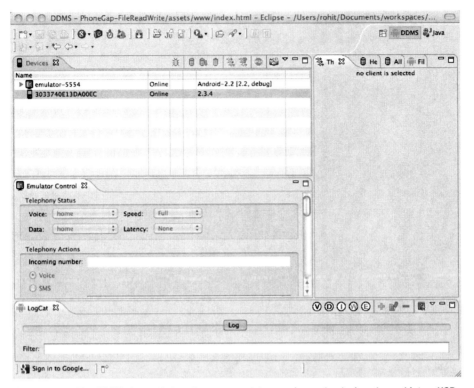

Figure 2–26. *The DDMS shows that we have an emulator running and a device plugged into a USB.*

2. When you click on the Run As of any Android application of your Android Project, you will see the screen in Figure 2–27. You are presented with this screen because you have both an emulator and a device available. Here Eclipse is asking you where to deploy the application. In case you only have a device plugged in and no emulator running, this screen will not be seen.

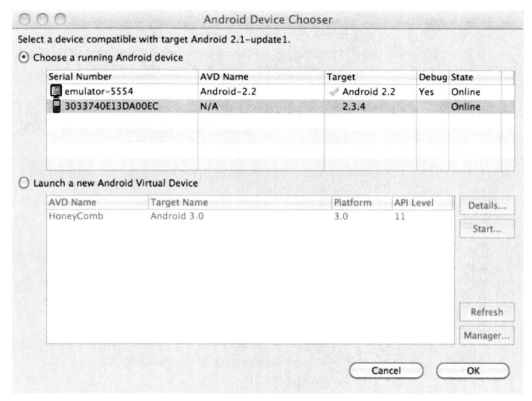

Figure 2–27. *In case there is more than one device or emulator, the DDMS will prompt the user to choose where to deploy the application.*

Exploring PhoneGap Features

This section explores more of the PhoneGap features.

Here is a short summary of features that PhoneGap supports:

1. The accelerometer API of PhoneGap enables the application to sense change in the device's orientation, therefore, it is able to act accordingly. This can be useful in creating applications that have a bubble level (making sure the phone is aligned horizontally to the ground). There is an option to fetch one reading of change in device orientation or to continuously receive the changes in device orientation.

2. The camera API of PhoneGap allows applications to retrieve a picture from either the camera (which is very useful for Facebook and Picasa applications) or fetch the images from already existing photo galleries.

3. The compass API of PhoneGap helps the applications know the bearing of the phone. This proves to be useful for map and navigation applications, since the map rotates as the user changes the bearing of the phone There is an option to fetch one reading of change in device heading or to continuously receive the changes in device heading.

4. The contacts API of PhoneGap is a way for applications to read and write contacts. Many social applications can benefit from syncing phone contacts with contacts on social channels.

5. The file API of PhoneGap allows applications to read, write, and list directories and file systems. This is handy if the application is planning to change the contents of a file in the file system of the phone. This API can also help write file explorer applications.

6. The geolocation API helps to retrieve the device's geolocation. This is good for many applications, including map-based applications, and applications like toursquare, where the user can check-in to a place by using their GPS location. There is an option to fetch one reading of change in device geo location or to continuously receive the changes in device geo location.

7. The media API allows applications to control the media sensors and applications on the device. This API allows applications to record and playback audio and video recordings.

8. The network API of PhoneGap provides the applications with the ability to see the state of the network. Instead of this state being just online and offline, this tells the application whether the device is on a 2G/3G/4G network or a Wi-Fi network. Such information often helps the application decide when to retrieve certain kinds of information.

9. The notification API allows applications to notify the user that something has occurred, by making a beep, vibration, or providing a visual alert.

10. The storage API of PhoneGap provides a built-in SQL database for the applications. An application can insert, retrieve, update, and delete data through SQL statements. Applications can query data in the database, and search for a specific e-mail in a locally stored list of e-mails.

PhoneGap Tutorials

Not all of the PhoneGap tutorials can be done on an Android emulator, so we will explain the following methods of examining the tutorials:

▓   Tutorials that can be done on an Android emulator.

▓   Tutorials that require an Android phone to work.

Emulator Examples

Fetching Device Information

PhoneGap allows the device information to be read programmatically. In order to do this you need to ensure that the PhoneGap framework has been loaded. Once the framework has been loaded, you can extract device information using JavaScript. All the properties of the Device Information are listed in Table 2–1.

Table 2–1. *Device Information Properties*

JavaScript Property	Description
device.name	Retrieves the device's model name.
device.phonegap	Retrieves the version of PhoneGap running on the device.
device.platform	Retrieves the device's operating system.
device.version	Retrieves the version of the device's operating system.
device.uuid	Retrieves the device's Universally Unique Identifier number.

The following code will give you access to the device's information. Refer to Figure 2–28 for the same.

```
<!DOCTYPE HTML>
<html>
  <head>

    <title>PhoneGap</title>

    <script type="text/javascript" src="phonegap-1.1.0.js"></script>

    <script type="text/javascript">

      /** Called when phonegap javascript is loaded */
```

```
      function onDeviceReady(){
        document.getElementById("deviceName").innerHTML
              = device.name;
        document.getElementById("version").innerHTML
              = device.phonegap;
        document.getElementById("mobilePlatform").innerHTML
              = device.platform;
        document.getElementById("platformVersion").innerHTML
              = device.version;
        document.getElementById("uuid").innerHTML
              = device.uuid;
      }

      /** Called when browser load this page*/
      function init(){
         document.addEventListener("deviceready", onDeviceReady, false);
      }

  </script>
</head>
<body onLoad="init()">
  <h1>Device Info</h1>
  <table border="1">
    <tr>
      <td>Device Name</td>
      <td id="deviceName"></td>
    </tr>
    <tr>
      <td>PhoneGap Version</td>
      <td id="version"></td>
    </tr>
    <tr>
      <td>Mobile Platform</td>
      <td id="mobilePlatform"></td>
    </tr>
    <tr>
      <td>Platform Version</td>
      <td id="platformVersion"></td>
    </tr>
    <tr>
      <td>UUID</td>
      <td id="uuid"></td>
    </tr>
  </table>
</body>
</html>
```

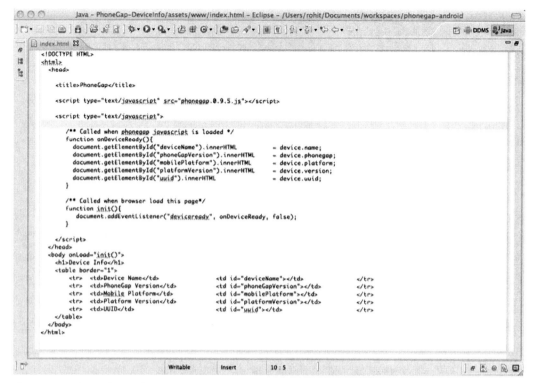

Figure 2–28. *PhoneGap device Info HTML source code*

When this code is run, the screen in Figure 2–29 should appear on your Android emulator.

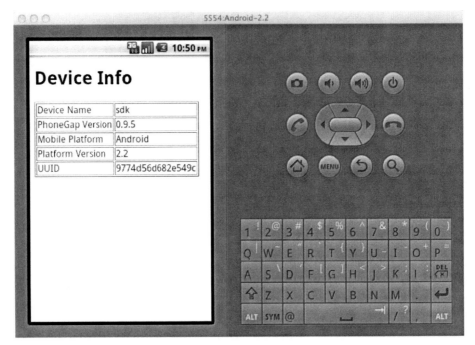

Figure 2–29. *Running the PhoneGap's device info on the emulator*

You can download the complete source for this example from
https://bitbucket.org/rohitghatol/apress-
phonegap/src/67848b004644/android/PhoneGap-DeviceInfo .

You can refer to the official documentation of Device API at
http://docs.phonegap.com/en/1.1.0/phonegap_device_device.md.html#Device.

Fetching the Device's Contacts

We are going to use PhoneGap to fetch the device's address book contact numbers. Before we do this on an Android emulator, we need to setup the Android emulator with some contact information.

1. Click on the dialer application icon 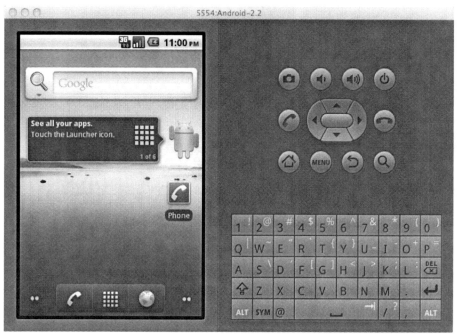, seen in Figure 2–30.

Figure 2–30. *Click on the PhoneGap application on the Android emulator.*

2. This will open up the dialer application. Click on the Contacts tab, seen in Figure 2–31.

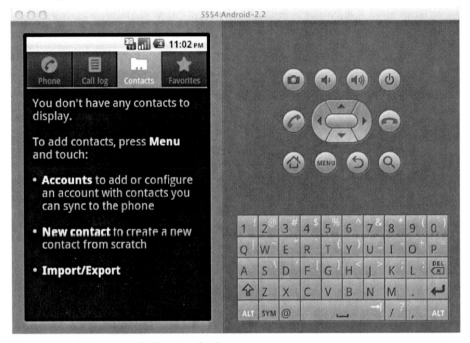

Figure 2–31. *Add a contact in Phone application*

3. Click Menu and choose New Contact. In the new contact add the first name, last name, and phone number. Refer to Figures 2–32 and 2–33.

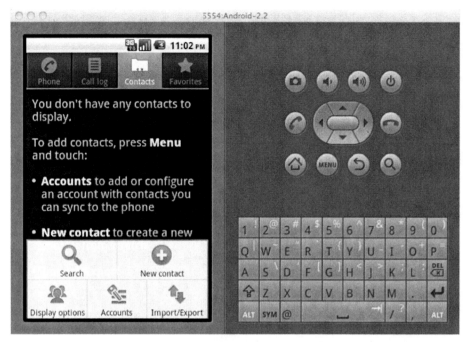

Figure 2–32. *Click Menu and click New Contact*

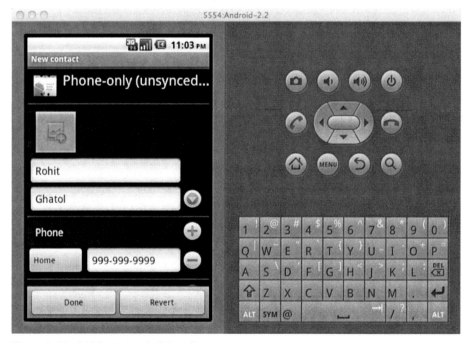

Figure 2–33. *Add Contact and click on Done.*

4. Once you have entered the needed value, click Done, and you should see one contact in your Contact List (as shown in Figure 2–34).

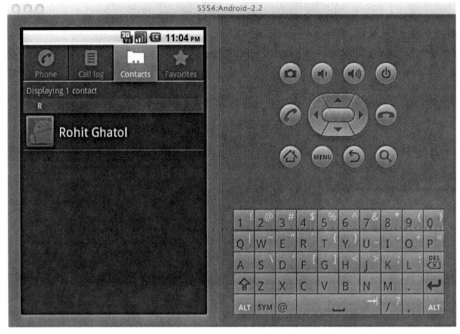

Figure 2–34. *Contact list*

In order to access the Contact List using PhoneGap, we need to use following API:

```
navigator.service.contacts.find(contactFields, contactSuccess, contactError,
contactfindOptions);
```

Table 2–2 provides a description of each argument.

Table 2–2. *Arguments for PhoneGap Contacts API*

Argument	Description	Example
contactField	Required argument. This is an array field of contacts that need to be returned.	["name","phoneNumbers"]

contactSuccess	A JavaScript callback function that gets a contact array as an argument.	function onSuccess(contacts){
		
		 }

contactError	A JavaScript callback function that gets an error as an argument.	function onError(error){
		
		 }

contactfindOptions	Options like filtering by name	var options = new ContactFindOptions() options.filter="Bob";

The following provides the steps to produce the complete code for fetching contacts:

1. Create a ContactFindOptions object. By making the options.filter equal to "" we are saying we want to fetch all contacts. If the options.filter was Bob, it would mean that we wanted to filter our search so that all of the results must contain the keyword Bob somewhere in the contact field.

```
var options = new ContactFindOptions();
options.filter="";
```

2. We need to define the contact fields that we want to fetch. When we search for contacts, a contact list appears. The contact itself is an associative array of contact fields. This means that if we specify that we want to fetch contacts only within the Name and Phone numbers contact field, we would only get that information back.

```
var fields = ["name","phoneNumbers"];
```

3. Define the call back method for success and failure. When we call the method Navigator.service.contacts.find(), we need to provide two callbacks. This is because the find() method is an asynchronous method.

```
function onSuccess(contacts) {
    for(var index=0;index<contacts.length;index++){
        var contact= contacts[index];
        var contactName = contact.name.formatted;
    }
}
function onError(error) {
}
navigator.service.contacts.find(fields, onSuccess, onError, options);
```

4. Use Android Linkify to allow us to dial numbers. While we list the contacts in the address book as a part of the HTML list (as seen in the following code),

```
<ul>
<li>Rohit Ghatol</li>
</ul>
```
we can make the listing more useful by using a link for the contact. This is depicted as follows:
```
<ul>
<li>
<a href="tel://999-999-9999">Rohit Ghatol</a>
</li>
</ul>
```

When a user clicks on the link Rohit Ghatol, Android reads the URL to be tel and opens the dialer to dial the number.

This is depicted in Figure 2–36 and Figure 2–37, which show how the application looks on Android Emulator.

Use the following code to apply what we have learned so far. Refer to Figure 2–35 for the index.html source code.

```html
<!DOCTYPE HTML>
<html>
  <head>
    <title>PhoneGap</title>
    <script type="text/javascript" src="phonegap-1.1.0.js"></script>
    <script type="text/javascript">
        /** Called when phonegap javascript is loaded */
        function onDeviceReady(){
                // find all contacts
                var options = new ContactFindOptions();
                options.filter="";
                var fields = ["phoneNumbers", "name"];
                navigator.service.contacts.find(fields, onSuccess, onError
                                               , options);
        }
        function onSuccess(contacts) {
             var ul = document.getElementById("list");
             for(var index=0;index<contacts.length;index++){
                 var name = contacts[index].name.formatted;
                 var phoneNumber = contacts[index].phoneNumbers[0].value;
                 var li = document.createElement('li');
                 li.innerHTML = "<a href=\"tel://"+phoneNumber+"\">"+name+"</a>";
                 ul.appendChild(li);
             }
        };
        function onError() {
             alert('onError!');
        };
        /** Called when browser load this page*/
        function init(){
            document.addEventListener("deviceready", onDeviceReady, false);
        }
    </script>
  </head>
  <body onLoad="init()">
    <h1>Contacts</h1>
    <ul id="list">
    </ul>
  </body>
</html>
```

```
Java - PhoneGap-Contacts/assets/www/index.html - Eclipse - /Users/rohit/Documents/workspaces/phonegap-android

index.html ⊠
<!DOCTYPE HTML>
<html>
    <head>
        <title>PhoneGap</title>
        <script type="text/javascript" src="phonegap.0.9.5.js"></script>
        <script type="text/javascript">
            /** Called when phonegap javascript is loaded */
            function onDeviceReady(){
                // find all contacts
                var options = new ContactFindOptions();
                options.filter="";
                var fields = ["phoneNumbers", "name"];
                navigator.service.contacts.find(fields, onSuccess, onError, options);
            }
            function onSuccess(contacts) {
                var ul = document.getElementById("list");
                for(var index=0;index<contacts.length;index++){
                    var name = contacts[index].name.formatted;
                    var phoneNumber = contacts[index].phoneNumbers[0].value;
                    var li = document.createElement('li');
                    li.innerHTML = "<a href=\"tel://"+phoneNumber+"\">"+name+"</a>";
                    ul.appendChild(li);
                }
            };
            function onError() {
                alert('onError!');
            };
            /** Called when browser load this page*/
            function init(){
                document.addEventListener("deviceready", onDeviceReady, false);
            }
        </script>
    </head>
    <body onLoad="init()">
        <h1>Contacts</h1>
        <ul id="list">
        </ul>
    </body>
</html>
```

Figure 2–35. *PhoneGap contact application HTML/JavaScript source code*

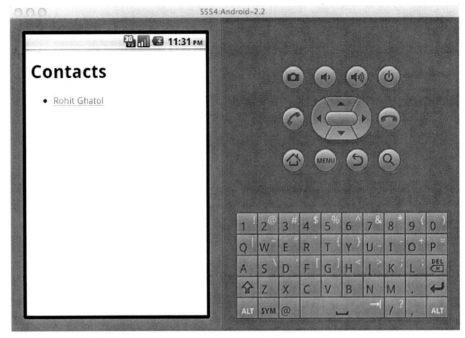

Figure 2–36. *Listing Contacts*

Figure 2–37. *Clicking on the Contact opens the phone dialer.*

You can download the complete source for this example from
`https://bitbucket.org/rohitghatol/apress-`
`phonegap/src/67848b004644/android/PhoneGap-Contacts.`

You can refer to the official documentation of Contacts API at
`http://docs.phonegap.com/en/1.1.0/phonegap_contacts_contacts.md.html#Contacts.`

Fetching the SD Card Listing

This section will explain how to list the SD card of an Android device. This section uses
a combination of W3C standards and a PhoneGap API.

There are two steps to listing the SD card of an Android device. These steps are as
follows:

1. We resolve the directory `file:///sdcard` and gain access to the directoryentry.

2. When we get access to the directoryentry, we can create a directoryreader from
    the directoryentry, and fetch the contents of the directory (SD card).

In Step 1, we call the following function:

```
window.resolveLocalFileSystemURI("file:///sdcard", onResolveSuccess, onError);
```

In response to the previous function, the onResolveSuccess callback will be activated.
The onResolveSuccess callback can be seen in the following code. Once we get access
to the fileEntry we create a directoryReader from it and call readEntries on that
directoryReader.

```
function onResolveSuccess(fileEntry){

var directoryReader = fileEntry.createReader();

directoryReader.readEntries(onSuccess,onError);
}
```

The onSuccess method is called when the path `file:///sdcard` is successfully resolved.
The onSuccess method is as follows:

```
function onSuccess(entries) {
document.getElementById("loading").innerHTML="";
var ul = document.getElementById("file-listing");
for(var index=0;index<entries.length;index++){
var li = document.createElement('li');
li.innerHTML = entries[index].name;
ul.appendChild(li);
    }
```

Deploy to the Simulator

The following code provides a complete example of how to deploy to the simulator:

```
<!DOCTYPE HTML>
<html>

    <head>
```

```
    <title>
        PhoneGap
    </title>
    <script type="text/javascript" src="phonegap-1.1.0.js">
    </script>
    <script type="text/javascript">
            /** Called when phonegap javascript is loaded */

    function onDeviceReady() {
        window.resolveLocalFileSystemURI("file:///sdcard",
            onResolveSuccess, onError);
    }

    function onResolveSuccess(fileEntry) {
        var directoryReader = fileEntry.createReader();
        directoryReader.readEntries(onSuccess, onError);
    }

    function onSuccess(entries) {
        document.getElementById("loading").innerHTML = "";
        var ul = document.getElementById("file-listing");
        for (var index = 0; index < entries.length; index++) {
            var li = document.createElement('li');
            li.innerHTML = entries[index].name;
            ul.appendChild(li);
        }
    }

    function onError(error) {
        alert('code: ' + error.code + '\n'
    + 'message: ' + error.message + '\n');
    }

     /** Called when browser load this page*/

    function init() {
        document.addEventListener("deviceready", onDeviceReady, false);
    }
    </script>
</head>

<body onLoad="init()">
    <h1>
        List SDCard Contents
    </h1>
    <ul id="file-listing">
    </ul>
    <div id="loading">
        Loading ..
    </div>
</body>

</html>
```

This code is illustrated in Figure 2–38.

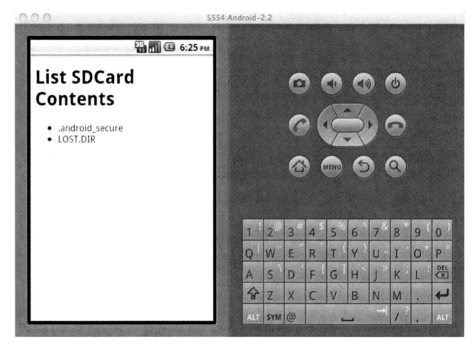

Figure 2–38. *Listing files on the SD Card*

You can download the complete source for this chapter from
https://bitbucket.org/rohitghatol/apress-
phonegap/src/67848b004644/android/PhoneGap-DirectoryListing.

You can refer to the official documentation of File API at
http://docs.phonegap.com/en/1.1.0/phonegap_file_file.md.html#File.

Writing and Reading to a File

This section will show you how to use the PhoneGap APIs to manipulate the file system.

> **NOTE:** PhoneGap is actually implementing and supporting the W3C's file system specs
> mentioned at http://www.w3.org/TR/file-system-api/.

Let's make ourselves familiar with some of the key concepts found in the file system's
API. These key concepts are:

1. LocalFileSystem

2. fileSystem

3. fileEntry

4. directoryEntry

LocalFileSystem

LocalFileSystem provides us with access to the local file system and its files and directories. There are two types of FileSystems:

1. **LocalFileSystem.PERSISTENT**: Data stored in a persistent file system should not be deleted by the UA, other than in response to a removal API call, without explicit authorization from the user.

2. **LocalFileSystem.TEMPORARY**: Data stored in a temporary file system may be deleted by the UA at its discretion, without application or user intervention.

The localFileSystem methods are accessed from the window object. This is done with the following:

1. window.requestFileSystem() – Used to gain access to the root file system.

2. window.resolveLocalFileSystemURI() – Used to directly gain access to either the FIleEntry or DIrectoryEntry Object glven that the URI is available for that directory or file.

fileSystem

The fileSystem represents the file system. It has two main properties:

1. name – choose between two options "PERSISTENT" or "TEMPORARY". Choose "PERSISTENT" if you want your files to persistent even when application is killed.

2. root – the Root Directory of the file system (DirectoryEntry).

You need to use the following API to get access to the fileSystem:

```
void requestFileSystem (
        short type,   // LocalFileSystem.PERSISTENT or
                        //LocalFileSystem.TEMPORARY
        long long size, //Size for TEMPORARY FS
        FileSystemCallback successCallback,  //Success Callback
        optional ErrorCallback errorCallback);  // Failure Callback
```

The code snippet for gaining access to the fileSystem is as follows:

```
window.requestFileSystem(LocalFileSystem.PERSISTENT
            ,0 //size
            ,function(fileSystem){ // success callbac
                alert("Got FileSystem "+fileSystem);
            },
            function(err){ //failure callback
                alert("Got Error requesting FileSystem");
            }
        );
```

You need to use the following API to gain access to either a fileEntry or directoryEntry object (objects which represent a file or directory):

```
void resolveLocalFileSystemURL (
            DOMString url,  //url of file or directory on
                                //filesystem
            EntryCallback successCallback,  //Success Callback
            optional ErrorCallback errorCallback); //FailureCallback
```

fileEntry

In order to manipulate a file you will need a fileEntry object. There are many ways to get a fileEntry, but given that you know the URI of a file, you can gain this object as follows:

```
window.resolveLocalFileSystemURL(
            "file:///sdcard/read-write.txt",
            function(fileEntry){
            },
            function(err){
            }
            );
```

In order to look up all of the properties and methods of a fileEntry, you need to have access to the appendix of the fileEntry. There are two ways of gaining this access:

1. createWriter(): Creates a FileWriter object that can be used to write to a file.

2. file(): Creates a File object containing file properties, including reading its content.

directoryEntry

In order to list files in a directory you will need a directoryEntry object. There are many ways to get a directoryEntry, but given that you know the URI of a directory, you can gain this object as follows:

```
window.resolveLocalFileSystemURL(
            "file:///sdcard/mydir/",
            function(directoryEntry){
            },
            function(err){
            }
            );
```

In order to look up all the properties and methods of a directoryEntry, you need to have access to the appendix of the directoryEntry. There is one method to gain this access: getFile(): Create File in a given directory,or Get File from a given directory.

Layout of Program

Our program for File Read and Write is simple. It contains a TextArea where we either read the file contents or we write the contents of a TextArea in a file. We use two buttons, Read and Write, to read/write from a file named read-write.txt.

Following is the code for the program.

```html
<!DOCTYPE HTML>
<html>
    <head>
        <title>PhoneGap</title>
        <script type="text/javascript" src="phonegap-1.1.0.js">
        </script>
        <script type="text/javascript">
            var filename = "read-write.txt";
            var filePath = "file:///sdcard/read-write.txt";
            var textarea = document.getElementById("textarea");
            /** Called when phonegap javascript is loaded */
            function onDeviceReady(){
                var readButton = document.getElementById("read");
                var writeButton = document.getElementById("write");

                readButton.addEventListener("click", readFile, false);
                writeButton.addEventListener("click", saveFile, false);

            }

            function readFile(){

                //Contents shown below

            }

            function saveFile(){

                //Contents shown below


            }



            /** Called when browser load this page*/
            function init(){
                document.addEventListener("deviceready", onDeviceReady,
                                    false);
            }
        </script>
    </head>
    <body onLoad="init()">
        <h1>Read Write File</h1>
        <table>
            <tr>
                <td colspan="2">
                    /sdcard/read-write.txt
                </td>
            </tr>
            <tr>
                <td colspan="2">
                    <textarea id="textarea" rows="10" cols="30">
                    </textarea>
                </td>
            </tr>
```

```
    <tr>
        <td>
            <button id="read">
                Read
            </button>
        </td>
        <td>
            <button id="write">
                Write
            </button>
        </td>
    </tr>
</table>
    </body>
</html>
```

This code when run is illustrated in Figure 2–39.

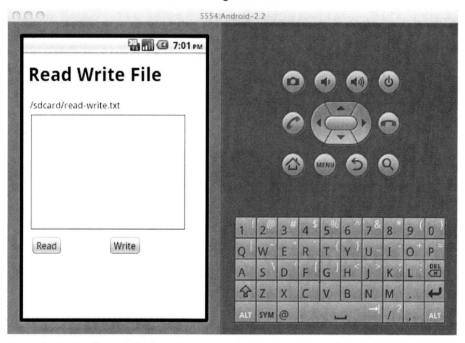

Figure 2–39. *Reading and Writing to Files.*

Now, let's implement the readFile() method to read the file and show its content in the TextArea. The steps are very simple:

Step 1: Resolve the URL `file:///sdcard/read-write.txt`.

Step 2: If resolved, create a reader using the fileEntry.file() method.

Step 3: If not resolved, show a message to the user telling the user that he needs to write the file before he can read it.. Refer to Figure 2–40.

```
function readFile(){

        window.resolveLocalFileSystemURI(    //filename to be read
            filePath,    //success callback
            function(fileEntry){
                fileEntry.file(
                    function(file){
                        var fileReader = new FileReader();
                        fileReader.onloadend =
                        function(evt){
                        document.getElementById("textarea").value
                      = evt.target.result;
                        };
                        fileReader.readAsText(file);
                            },
                        function(error){
                            alert("Got error while reading
"+filePath);
                        })
            },    //error callback
            function(error){
                alert(filename + " not present, please add content and click
                    Save first");
            }
        );

}
```

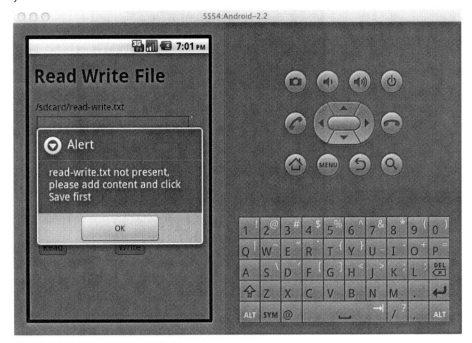

Figure 2–40. *Unable to read as no file has been written.*

Let's implement the writeFile() method to read the text in the TextArea, and write it in file:///sdcard/read-write.txt.

Step 1: Get Access to the fileSystem root.

Step 2: From the fileSystem root DirectoryEntry, create file read-write.txt, if it does not already exist.

Step 3: Create fileWriter and write the contents of the TextArea to the file. This is shown in the following code:

```
function saveFile() {

    window.requestFileSystem(
    LocalFileSystem.PERSISTENT, 0,
    //Success Callback


    function (fileSystem) {
        var sdcardEntry = fileSystem.root;
        sdcardEntry.getFile(
        filename,
        //Flag telling create file
        {
            create: true
        },
        //Success callbacks


        function (fileEntry) {
            fileEntry.createWriter(

            function (fileWriter) {
                fileWriter.onwrite = function (evt) {
                    alert("Write was successful!");
                    document.getElementById("textarea").value = "";
                };
                fileWriter.write(document.getElementById("textarea").value);
            },
            //Error callback


            function (error) {
                alert("Failed to get a file writer for " + filename);
            });



        },
        //Error Callback


        function (error) {
            alert("Got error while reading " + filename + " " + error);
        });

    }, function (error) {
```

```
    alert("Got Error while gaining access to file system");
});
```

```
}
```

When the user types text in the TextArea and hits Write, the contents are written to the file and an alert message is shown, saying Write was successful. You will notice the program clears the TextArea when you hit Write. Refer to Figure 2–41 to see the message when file writing is successful.

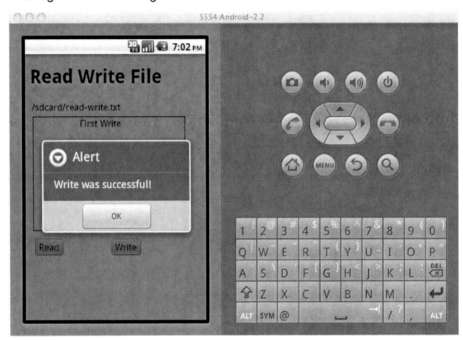

Figure 2–41. *Write was successful*

Now, try to read the content that you wrote to the file, by hitting the Read button. The content of what you wrote should appear in the TextArea. Refer to Figure 2–42 to see how the text area is filled, when reading is successful.

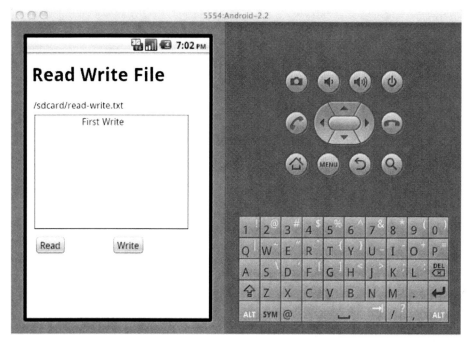

Figure 2–42. *Read was successful*

You can download the complete source for this example from
`https://bitbucket.org/rohitghatol/apress-`
`phonegap/src/62c45e339662/android/PhoneGap-FileReadWrite.`

You can refer to the official documentation of File API at
`http://docs.phonegap.com/en/1.1.0/phonegap_file_file.md.html#File.`

Writing and Reading from the Database

This section will explain how to read, store, and manipulate the database. We will use
the example of storing a list of contacts (firstName and lastName) in the database, and
then read it, and delete the entry.

There are three tables: one for the header columns, one for the content, and one to allow
the user to add an entry to the list of contacts.

Following is the index.html code for reading and writing from the database:

```
<!DOCTYPE HTML>
<html>
    <head>
        <title>PhoneGap DB</title>
        <script type="text/javascript" src="phonegap-1.1.0.js">
        </script>
        <script type="text/javascript">
```

```
        var firstNameBox = null;
            var lastNameBox = null;
        var db = null;
        var dataTable = null;
        /** Called when phonegap javascript is loaded*/
        function onDeviceReady(){
            //Contents will be shown below
            }
        /** Called when browser load this page*/
        function init(){
            document.addEventListener("deviceready",
                        onDeviceReady, false);
        }
    </script>
    <style>
        td {
            width: 100px;
        }

        input {
            width: 100px;
        }
    </style>
</head>
<body onLoad="init()">
    <h3>Read Write DB</h3>
    <table border="1">
        <tr>
            <td>
                <b>First Name</b>
            </td>
            <td>
                <b>Last Name</b>
            </td>
            <td>
                <b>Action</b>
            </td>
        </tr>
    </table>
    <table id="data-table">
    </table>
    <table>
        <tr>
            <td>
                <input id="firstName" type="text">
                </input>
            </td>
            <td>
                <input id="lastName" type="text">
                </input>
            </td>
            <td>
                <button id="add">
                    Add
                </button>
            </td>
        </tr>
```

```
        </table>
    </body>
</html>
```

If you run this program you should see a layout similar to the one shown in Figure 2–43

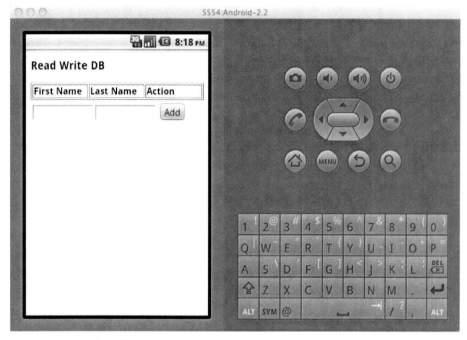

Figure 2–43. *Read and Write to and from the database*

Now, let's add some code to read, write, and delete contact entries from the database. The first step is to get access to the database object. This is done as follows:

```
var firstNameBox = null;
var lastNameBox = null;
var db = null;
var dataTable = null;
/** Called when phonegap javascript is loaded */
function onDeviceReady(){
        var addButton = document.getElementById("add");
        firstNameBox = document.getElementById("firstName");
        lastNameBox = document.getElementById("lastName");
        dataTable = document.getElementById("data-table");

        db = window.openDatabase("contactDB", "1.0", "Contact
                Database", 1000000);
                //name,version,display name, size

}
```

Let's focus on the API to create the database object. This is done as follows:

```
window.openDatabase(
        databaseName,
        versionNumber,
        displayName,
        sizeInBytes);
```

Now let's see the code (refer to addButton.addEventListener below) that explains the process of clicking on the Add button, and adding an entry to the contacts table in the database.

In order to add entry to contacts table we will use the method transaction() on the database object. Here is the API of this method:

```
db.transaction(
      function(tx){   //Function to execute the sql statements
            //Use tx to execute sql statements
      },
      function(err){  // Error callback
            //Use err.code and err.message to understand the error
      },
      function(){  //Success callback
            // Update the UI, log a message
      }
      );
```

We want to add the new entry when the user clicks on the button. Hence, our add functionality is inside the click event listener of the Add button. This is seen in the following code:

```
addButton.addEventListener(
      "click",
      function(){

            db.transaction(
                    //function sql statements
                    function (tx){
                          ensureTableExists(tx);
                          var firstName = firstNameBox.value;
                          var lastName = lastNameBox.value;

                          var sql = 'INSERT INTO Contacts
                    ( firstName, lastName ) VALUES
                    ("' + firstName + '","' + lastName + '")';

                          tx.executeSql(sql);

                    },
                    //error callback
                    function (err){
                          alert("error callback "+err.code);

                    },
                    //success callback
                    function (){
                          loadFromDB();
                    }
```

```
            );

    },false);
function ensureTableExists(tx){
    tx.executeSql('CREATE TABLE IF NOT EXISTS Contacts (id
                INTEGER PRIMARY KEY, firstName,lastName)');

}
```

We are using a method named ensureTableExists(tx), which ensures that we have the database table before we do a DB CRUD Operation in the database.

We are using the SQL insert statement in our database transaction's SQL function in order to make an entry in the database.

NOTE: The primary key in SQLite is auto incremented by the default, hence, we only write the firstName and lastName, and the database actually increments the ID.

Once the database has added the operation successfully, we call the loadFromDB() method to populate the HTML table from the database table.

Refer Figure 2–44 to see how the UI Screen looks like for Add Contact functionality.

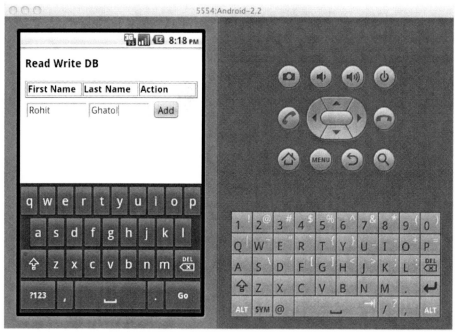

Figure 2–44. *Adding an entry*

Now, let's look into the loadFromDB() method. Here we are using the same tx.executeSql, but in this case, we expect to get some different results, so we use the following version of tx.executeSql:

```
tx.executeSql(
        sqlStatement,
        options,
        successCallbackWithResultSet,
        errorCallback);
```

The successCallbackWithResultSet is a function that receives two things:

1. tx

2. resultset

```
function loadFromDB(){

db.transaction(
        //function sql statements
        function (tx){
                ensureTableExists(tx);
                tx.executeSql('SELECT * FROM Contacts',
                        [],
                        //success callback
                                        function(tx, results){
                                        var htmlStr="";
                                                for(var
index=0;index<results.rows.length;index++){
                                                        var item =
results.rows.item(index);

                        htmlStr = htmlStr +"<tr><td>"+
                                item.firstName+"</td><td>"
                                +item.lastName
                                +"</td><td><button
                                onclick=\"deleteEntry('"
                                +item.id+
                                "');\">X</button></td></tr>";

                                }

                                dataTable.innerHTML=htmlStr;
                        },
                        //error callback
                        function(err){
                                alert("Unable to fetch result from Contacts
                                Table");
                        }
                );

        },
        //error callback
        function (err){
                alert("error callback "+err.code+" "+err.message);

        },
        //success callback
```

```
        function (){
                firstNameBox.value="";
                lastNameBox.value="";
        });

}
```

When we run the code now, we can see the previously added entries in the HTML table.

Refer Figure 2–45, it shows previously added rows in the html table.

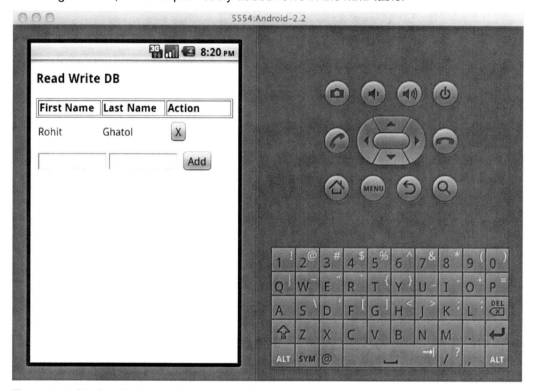

Figure 2–45. *Reading all of the entries from the database*

The last thing to do is to delete the entry when the user clicks on the X button. When we populated the HTML table, we defined the button in HTML as follows:

```
<button onclick="deleteEntry('"+item.id+"');'>X</button>
```

Refer to Figure 2–46 to see how this looks in the HTML.

When the user clicks on the X button, the deleteEntry function is called, passing the primary key of the entry to be deleted. Here, we are using the Delete SQL statement to delete the primary key.

```
function deleteEntry(id){
        db.transaction(
                //function sql statements
                function (tx){
```

```
                ensureTableExists(tx);

                  tx.executeSql('Delete FROM Contacts where id='+id);

            },
            //error callback
            function (err){
                  alert("error callback "+err.code+" "+err.message);

            },
            //success callback
            function (err){
                  loadFromDB();
            }
      );

}
```

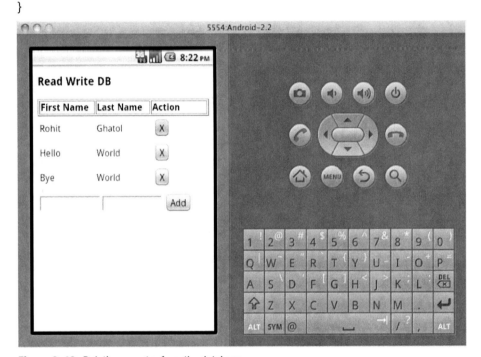

Figure 2–46. *Deleting an entry from the database*

Now when the user clicks on the X button, the entry is deleted and a call is given to loadFromDB(), which refreshes the HTML table from the database table.

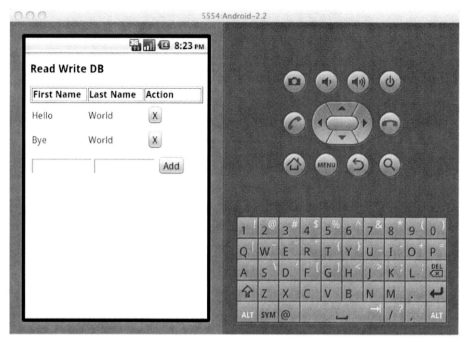

Figure 2–47. *Delete entry is reflected*

The complete index.html source is listed here:

```
<!DOCTYPE HTML>
<html>

    <head>
        <title>
            PhoneGap DB
        </title>
        <script type="text/javascript" src="phonegap-1.1.0.js">

        </script>
        <script type="text/javascript">
                        var firstNameBox = null;
                var lastNameBox = null;
                var db = null;
                var dataTable = null; /** Called when phonegap javascript is loaded */

                function onDeviceReady() {
                    var addButton = document.getElementById("add");
                    firstNameBox = document.getElementById("firstName");
                    lastNameBox = document.getElementById("lastName");
                    dataTable = document.getElementById("data-table");

                    db = window.openDatabase("contactDB",
                    "1.0",
                     "Contact Database",
                    1000000); //name,version,display name, size
                    addButton.addEventListener("click", function() {
```

```
                    db.transaction(
                    //function sql statements

                    function(tx) {
                        ensureTableExists(tx);
                        var firstName = firstNameBox.value;
                        var lastName = lastNameBox.value;

                        var sql = 'INSERT INTO Contacts (firstName, lastName)
VALUES
                        ("' + firstName + '","' + lastName + '")';
                        tx.executeSql(sql);

                    },
                    //error callback

                    function(err) {
                        alert("error callback " + err.code);

                    },
                    //success callback

                    function(err) {
                        //alert("success callback "+err.code);
                        loadFromDB();
                    });

                }, false);
                loadFromDB();

            }

        function loadFromDB() {

            db.transaction(
            //function sql statements

            function(tx) {
                ensureTableExists(tx);
                tx.executeSql('SELECT * FROM Contacts', [], function(tx,
results) {

                    var htmlStr = "";
                    for (var index = 0; index < results.rows.length;
index++) {

                        var item = results.rows.item(index);
                        htmlStr = htmlStr
    + "<tr><td>"
    + item.firstName
    + "</td><td>"
    + item.lastName
    + "</td><td><button onclick=\"deleteEntry('"
    + item.id
    + "');\">X</button></td></tr>";

                    }
```

```
                    dataTable.innerHTML = htmlStr;
            }, function(err) {
                    alert("Unable to fetch result from Contacts Table");
            });

        },
        //error callback

        function(err) {
            alert("error callback " + err.code + " " + err.message);

        },
        //success callback

        function() {
            firstNameBox.value = "";
            lastNameBox.value = "";

        });

}

function deleteEntry(id) {
    db.transaction(
    //function sql statements

    function(tx) {
        ensureTableExists(tx);
        tx.executeSql('Delete FROM Contacts where id=' + id);

    },
    //error callback

    function(err) {
        alert("error callback " + err.code + " " + err.message);

    },
    //success callback

    function(err) {
        //alert("success callback ");
        loadFromDB();

    });


}

function ensureTableExists(tx) {
    tx.executeSql('CREATE TABLE IF NOT EXISTS Contacts
    (id INTEGER PRIMARY KEY, firstName,lastName)');

} /** Called when browser load this page*/

function init() {
    document.addEventListener("deviceready",
```

```
                                            onDeviceReady, false);
                    }
        </script>
        <style>
            td { width: 100px; } input { width: 100px; }
        </style>
    </head>

    <body onLoad="init()">
        <h3>
            Read Write DB
        </h3>
        <table border="1">
            <tr>
                <td>
                    <b>
                        First Name
                    </b>
                </td>
                <td>
                    <b>
                        Last Name
                    </b>
                </td>
                <td>
                    <b>
                        Action
                    </b>
                </td>
            </tr>
        </table>
        <table id="data-table">
        </table>
        <table>
            <tr>
                <td>
                    <input id="firstName" type="text">
                    </input>
                </td>
                <td>
                    <input id="lastName" type="text">
                    </input>
                </td>
                <td>
                    <button id="add">
                        Add
                    </button>
                </td>
            </tr>
        </table>
    </body>

</html>
```

You can download the complete source for this example from https://bitbucket.org/rohitghatol/apress-phonegap/src/62c45e339662/android/PhoneGap-DB.

You can refer to the official documentation of Storage API at
`http://docs.phonegap.com/en/1.1.0/phonegap_storage_storage.md.html#Storage`.

Fetching Details about a Cellular Device or a Wi-Fi Network

Often a Mobile application needs to connect to some server in order to fetch certain data. A modern day smartphone may be downloading the data by using either a 3G/4G network or a Wi-FI network. A good application would respect this distinction and download certain kinds of data when on a 3G/4G network, and only download heavy data when the smartphone is on a Wi-Fi network.

This section will explain how to use PhoneGap to find out which kind of network the smartphone is using.

We will need to use the following API:

```
navigator.network.connection.type
```

This returns the type of connection. The values for different connection types are mentioned below:

```
Connection.UNKNOWN = "unknown";
Connection.ETHERNET = "ethernet";
Connection.WIFI = "wifi";
Connection.CELL_2G = "2g";
Connection.CELL_3G = "3g";
Connection.CELL_4G = "4g";
Connection.NONE = "none";
```

The complete source code for the index.html is listed below:

```
<!DOCTYPE HTML>
<html>
    <head>
        <title>PhoneGap DB</title>
        <script type="text/javascript" src="phonegap-1.1.0.js">
        </script>
        <script type="text/javascript">


            /** Called when phonegap javascript is loaded */
            function onDeviceReady(){
                fetchNetworkConnectionInfo();

            }

            function fetchNetworkConnectionInfo(){

                var networkType = navigator.network.connection.type;

                var networkTypes = {};

                networkTypes[Connection.NONE]      = 'No network connection';
                networkTypes[Connection.UNKNOWN]
                   = 'Unable to identify Network Connection Type';
                networkTypes[Connection.CELL_2G]
```

```
                            = 'Network Connection is of type 2G';
                    networkTypes[Connection.CELL_3G]
                            = 'Network Connection is of type 3G';
                    networkTypes[Connection.CELL_4G]
                            = 'Network Connection is of type 4G';
                    networkTypes[Connection.WIFI]
                            = 'Network Connection is of type WiFi';
                    networkTypes[Connection.ETHERNET]
                            = 'Network Connection is of type Ethernet';


                    document.getElementById("network-status").innerHTML
                            = networkTypes[networkType];


            }

            /** Called when browser load this page*/
            function init(){
                document.addEventListener("deviceready",
                                    onDeviceReady, false);
            }
        </script>
    </head>
    <body onLoad="init()">
        <h3>Phone Network Info</h3>
        <div id="network-status">
        </div>
    </body>
</html>
```

When this source is run it would show the network info as depicted in Figure 2–48.

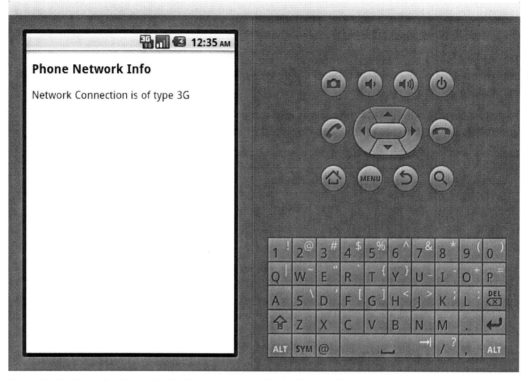

Figure 2–48. *PhoneGap Network Info showing Network Connection is of Type 3G*

You can download the complete source for this example from
https://bitbucket.org/rohitghatol/apress-
phonegap/src/67848b004644/android/PhoneGap-Network.

You can refer to the official documentation of Connection API at
http://docs.phonegap.com/en/1.1.0/phonegap_connection_connection.md.html#Connec
tion.

Device Examples

Given the nature of the features, the following examples can only be run on real Android
devices. The Android emulator does not have support for the following features.

Fetching the Geolocation

In this example we will try to fetch the geolocation of the device. In order to do this, we
will use the navigator.geolocation API. This API is an asynchronous API, which means
that once we request the API for the geolocation, the API will inform the called program,
and use one of the two callbacks registered with the API.

The API invoked is as follows:

```
navigator.geolocation.getCurrentPosition(onSuccessCallback, onErrorCallback);
```

The API calls the onSuccessCallback function when the API is able to fetch the GPS coordinates. The API will call onErrorCallback when there is a problem fetching the GPS coordinates.

The onSuccessCallback will get an argument named position. The position will contain the details about the geolocation as depicted in Table 2–3.

Table 2–3. *Position Object of PhoneGap GeoLocation API Explained*

Position result	Description
position.coords.latitude	Latitude in decimal degrees
position.coords.longitude	Longitude in decimal degrees
position.coords.altitude	Altitude in decimal degrees
position.coords.accuracy	Accuracy for Latitude and Longitude in meters
position.coords.altitudeAccuracy	Accuracy for Altitude in meters
position.coords.heading	Direction of travel in terms of degrees from the true north in a clockwise direction
position.coords.speed	Speed of travel in terms of meters per second
position.timestamp	Creation timestamp of the GPS coordinates

The complete code example of this is seen as follows:

```
<!DOCTYPE HTML>
<html>
  <head>
    <title>PhoneGap</title>
    <script type="text/javascript" src="phonegap-1.1.0.js"></script>
    <script type="text/javascript">
      /** Called when phonegap javascript is loaded */
      function onDeviceReady(){
          navigator.geolocation.getCurrentPosition(onSuccess, onError);
      }

      function onSuccess(position) {
          document.getElementById('latitude').innerHTML = position.coords.latitude;
          document.getElementById('longitude').innerHTML = position.coords.longitude;
          document.getElementById('altitude').innerHTML = position.coords.altitude;
```

```
            document.getElementById('timestamp').innerHTML = new Date(position.timestamp);
       }

     function onError(error) {
            alert('code: '    + error.code    + '\n' +
                           'message: ' + error.message + '\n');
       }

     /** Called when browser load this page*/
     function init(){
            document.addEventListener("deviceready", onDeviceReady, false);
       }
   </script>
 </head>
 <body onLoad="init()">
   <h1>GeoLocation</h1>
   <table border="1">
     <tr>  <td>Latitue</td>       <td id="latitude"></td>     <tr>
     <tr>  <td>Longitude</td>     <td id="longitude"></td>    <tr>
     <tr>  <td>Altitude</td>      <td id="altitude"></td>     <tr>
     <tr>  <td>Timestamp</td>     <td id="timestamp"></td>    <tr>
   </table>
   </ul>
 </body>
</html>
```

This code is illustrated in Figure 2–49.

Figure 2–49. *Running example of PhoneGap Geo location*

You can download the complete source for this example from
`https://bitbucket.org/rohitghatol/apress-`
`phonegap/src/67848b004644/android/PhoneGap-GeoLocation`.

You can refer to the official documentation of Geolocation API at
`http://docs.phonegap.com/en/1.1.0/phonegap_geolocation_geolocation.md.html#Geol`
`ocation`.

Fetching the Accelerometer

In this example, we will try to watch the accelerometer readings from the device. The accelerometer feature in modern day smart phones provides the user with the direction of their motions in x, y, and z coordinates.

The API invoked is as follows:

```
navigator.accelerometer.watchAcceleration(onSuccessCallback,
onErrorCallback,accelerometerOptions);
```

This API will keep monitoring the accelerometer readings and call onSuccessCallback at a predefined interval until the navigator.accelerometer.clearwatch() is called. The interval is defined in the accelerometerOptions, by providing the value in the form of {"frequency":"3000"}. The unit of interval is milliseconds. If the accelerometerOptions does not provide the default interval of 1000, milliseconds is used.

As mentioned in previous examples, the API will call onErrorCallback when there is a problem fetching the GPS coordinates.

The onSuccessCallback will get an argument named acceleration. The acceleration will contain the details about the device motion as depicted in the Table 2–4.

Table 2–4. *Acceleration Object of Accelerometer Explained*

Accleration	Description
x	Motion around the x-axis. The value is between 0-1.
y	Motion around the y-axis. The value is between 0-1.
z	Motion around the z-axis. The value is between 0-1.
timestamp	Creation timestamp of device motion.

The complete example of the code is as follows:

```html
<!DOCTYPE HTML>
<html>
  <head>
    <title>PhoneGap</title>
    <script type="text/javascript" src="phonegap-1.1.0.js"></script>
    <script type="text/javascript">
      /** Called when phonegap javascript is loaded */
```

```
        function onDeviceReady(){
            var options = { frequency: 1000 };   // Update every 1 seconds
              navigator.accelerometer.watchAcceleration(onSuccess,
                                      onError,options);
        }

        function onSuccess(acceleration) {
            document.getElementById('x').innerHTML = acceleration.x;
            document.getElementById('y').innerHTML = acceleration.y;
            document.getElementById('z').innerHTML = acceleration.z;
            document.getElementById('timestamp').innerHTML
            = acceleration.timestamp;
    }

       function onError(error) {
          alert('code: '    + error.code    + '\n' +
                        'message: ' + error.message + '\n');
          }

   /** Called when browser load this page*/
   function init(){
      document.addEventListener("deviceready", onDeviceReady, false);
   }
   </script>
 </head>
 <body onLoad="init()">
   <h1>Accelerometer</h1>
   <table border="1">
     <tr>  <td>X</td>          <td id="x"></td>           <tr>
     <tr>  <td>Y</td>          <td id="y"></td>           <tr>
     <tr>  <td>Z</td>          <td id="z"></td>           <tr>
     <tr>  <td>Timestamp</td>  <td id="timestamp"></td>   <tr>
   </table>
   </ul>
 </body>
</html>
```

This code is illustrated in Figure 2–50.

Figure 2–50. *PhoneGap Accelerometer example*

 A pictorial version of the same accelerometer is found in Figure 2–51. The accelerometer can be used to check the level of a surface by placing the phone flat on that surface.

You can find the graphics for this example at http://code.google.com/p/beginingphonegap/downloads/list.

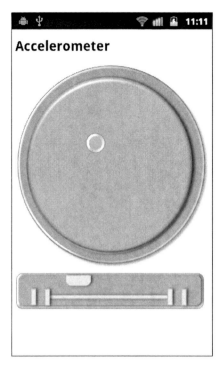

Figure 2–51. *Bubble application build using PhoneGap Accelerometer API*

There were four images used in the previous example. These images varied from a circular bubble to an oval bubble. The complete example of the code is seen as follows:

```
<!DOCTYPE HTML>
<html>
  <head>
    <title>PhoneGap</title>
    <script type="text/javascript" src="phonegap-1.1.0.js"></script>
    <script type="text/javascript">
      /** Called when phonegap javascript is loaded */
     function onDeviceReady(){
          var options = { frequency: 0100 };  // Update every 1 seconds
          navigator.accelerometer.watchAcceleration(onSuccess,
                                                    onError,options);
          }

    function onSuccess(acceleration) {
         moveX(acceleration);
         moveXY(acceleration);
     }

    function moveXY(acceleration){

          var xyBase = document.getElementById("x-y-base");
          var circle = document.getElementById("circle");
          var position = getPos(xyBase);
          var adjustX = 20;
```

```
            var adjustY = 20;
            var radius = 160;
            var left = position.x;
            var top  = position.y;
            var width  = xyBase.clientWidth;
            var height = xyBase.clientHeight;

            var centerX = left + width/2 - adjustX;
            var centerY = top + height/2 - adjustY;
            centerY = centerY - (radius * acceleration.y *  1.2) /10;
            centerX = centerX - (radius * acceleration.x * -1.2) /10;

            circle.style.left=centerX+"px";
            circle.style.top=centerY+"px";

    }
    function moveX(acceleration){
        //FIXME Move local variables to make them global
        var xBase = document.getElementById("x-base");
          var oval = document.getElementById("oval");
          var basePosition = getPos(xBase);

          var ovalLeft = basePosition.x + (xBase.clientWidth/2) -
              (xBase.clientWidth * acceleration.x * -1)/10;

          if( ( ovalLeft + oval.clientWidth )>
              (xBase.clientWidth+basePosition.x) ){
                ovalLeft = xBase.clientWidth + basePosition.x -
                    oval.clientWidth;
          }
          if (ovalLeft < basePosition.x){
                ovalLeft = basePosition.x;
          }
          oval.style.left=ovalLeft+"px";
    }

    function onError(error) {
          alert('code: '    + error.code   + '\n' +
                        'message: ' + error.message + '\n');
    }

      /** Called when browser load this page*/
      function init(){
         document.addEventListener("deviceready", onDeviceReady, false);
      }
      function getPos(el) {
            var position = {};
            if (document.getBoxObjectFor) {
                  var bo = document.getBoxObjectFor(el);
                  position.x = bo.x;
                  position.y = bo.y;
          }
          else {
                var rect = el.getBoundingClientRect();
                position.x = rect.left;
                position.y = rect.top;
          }
```

```
            return position;
        }
    </script>
</head>
<body onLoad="init()">
    <h1>Accelerometer</h1>

    <div id="horizontal-bubble">
        <img id="circle" src="accelerometer-circle-bubble.png"
            style="position:absolute"></img>
        <img id="x-y-base" src="x-y-accelerator-base.png"></img>
    </div>

    <div id="vertical-bubble">
        <img id="x-base" src="z-accelerator-base.png"></img>
        <img id="oval" src="accelerometer-circle-oval.png"
            style="position:absolute;left:0px"></img>
    </div>

</body>
</html>
```

You can download the complete source for this example from
https://bitbucket.org/rohitghatol/apress-
phonegap/src/67848b004644/android/PhoneGap-Accelerometer-Image .

You can refer to the official documentation of Accelerometer API at
http://docs.phonegap.com/en/1.1.0/phonegap_accelerometer_accelerometer.md.html#
Accelerometer.

Fetching Compass Bearings

Let's move to an application that has similar functionality as an accelerometer; a
compass. Like an accelerometer, a compass provides device motion in the degrees of
the x, y, and z axes. A compass provides device's heading in degrees from true north in
clock wise direction.

You can find the graphics for this example at
http://code.google.com/p/beginingphonegap/.

The API invoked is as follows:

```
navigator.compass.watchHeading (onSuccessCallback, onErrorCallback,compassOptions);
```

This API will keep monitoring the direction that the compass is heading in and call
onSuccessCallback at a predefined interval until navigator.compass.clearwatch() is
called. The interval is defined in that compassOptions, and provides a value in the form
of {"frequency":"3000"}. The unit of interval is milliseconds. If the compassOptions is not
provided, the default interval of 1000 milliseconds is used.

As mentioned in previous examples, the API will call onErrorCallback when there is a
problem fetching the GPS coordinates.

The onSuccessCallback will get an argument named heading. A heading is a degree between 0 and 360, and it is measured in a clock wise direction from the true north.

The complete example is as follows. This example uses CSS3 to visually show a compass. In order to do this, we used a compass pointer image. This image is shown in Figure 2–52.

Figure 2–52. *Image of Compass used in the PhoneGap example*

First we register our onSuccess method with navigator.compass.watchHeading() method. When our onsuccess method gets called, we change the direction where the compass image points using css3 rotate transformation. This is a visual compass application.

The complete index.html for this application is mentioned here:

```
<!DOCTYPE HTML>
<html>
  <head>
    <title>PhoneGap</title>
    <script type="text/javascript" src="phonegap-1.1.0.js"></script>
    <script type="text/javascript">
      /** Called when phonegap javascript is loaded */
          function onDeviceReady(){
              var button = document.getElementById("capture");
              var compassOptions = { frequency: 1000 };
              navigator.compass.watchHeading(onSuccess, onError,
                                  compassOptions);
          };

          function onSuccess(heading) {
              var image = document.getElementById('compass');
              var headingDiv =
              document.getElementById('compassHeading');
              headingDiv.innerHTML=heading;
              var reverseHeading = 360 - heading;
              image.style.webkitTransform =
              "rotate("+reverseHeading+"deg)";
          }

      function onError(error) {
          alert('code: '    + error.code    + '\n' +'message: ' + error.message +
'\n');
          }
```

```
     /** Called when browser load this page*/
        function init(){
            document.addEventListener("deviceready", onDeviceReady,
                                        false);
        }
   </script>
</head>
<body onLoad="init()">
   <h1>Compass</h1>
   <table>
       <tr>
               <td>Compass Heading</td>
               <td>
                       <div id="compassHeading">....</div>
               </td>
               <td>Degrees</td>
       </tr>
   </table>

   <img id="compass" src="compass.png"
       style="width:400px;height:400px;margin-left:auto;margin-
       right:auto;auto;display:block"></img>

 </body>
</html>
```

This code is illustrated in Figure 2–53.

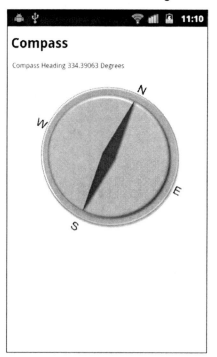

Figure 2–53. *PhoneGap Visual Compass application*

The Figure shown in 2–53 can be downloaded from this url - `http://beginingphonegap.googlecode.com/files/compass.png`.

You can download the complete source for this example from `https://bitbucket.org/rohitghatol/apress-phonegap/src/67848b004644/android/PhoneGap-Compass`.

You can refer to the official documentation of Compass API at `http://docs.phonegap.com/en/1.1.0/phonegap_compass_compass.md.html#Compass`.

Capturing an Image from the Camera

The last section in this chapter is about capturing images from the camera. This is a cool feature, which really adds a lot of value to HTML-based applications. Let's see how we can use this feature.

The API invoked is as follows:

```
navigator.camera.getPicture (onSuccessCallback, onErrorCallback,cameraOptions);
```

While there are many options present in the cameraOptions, we will focus on a single property named quality. The cameraOption would look like {"quality":75}.

With the above cameraOptions, the onSuccess() method will get an image of a base64 encoded binary.

The complete example of the Camera application is mentioned here:

```
<!DOCTYPE HTML>
<html>
  <head>
    <title>PhoneGap</title>
    <script type="text/javascript" src="phonegap-1.1.0.js"></script>
    <script type="text/javascript">
      /** Called when phonegap javascript is loaded */
function onDeviceReady(){

    var button = document.getElementById("capture");

  button.addEventListener("click",captureImage,false);

          }

function captureImage(){

    var cameraOptions = { quality: 50 };

  navigator.camera.getPicture( onSuccess, onError,

            cameraOptions );

};


function onSuccess(imageData) {

    var image = document.getElementById('cameraImage');
    image.src = "data:image/jpeg;base64," + imageData;
```

```
            }
function onError(error) {
             alert('code: '    + error.code    + '\n' +'message: ' + error.message +
'               \n');
}


/** Called when browser load this page*/
function init(){
    document.addEventListener("deviceready", onDeviceReady,

        false);

}

    </script>
  </head>
  <body onLoad="init()">
    <h1>Camera</h1>
    <button id="capture" >Capture Image</button>

    <img id="cameraImage"></img>

  </body>
</html>
```

This code is illustrated in Figure 2–54 and Figure 2–55.

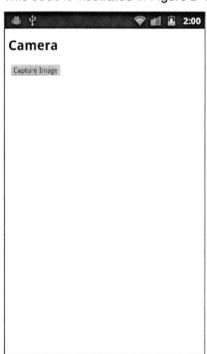

Figure 2–54. *PhoneGap Camera application showing button to Capture Image*

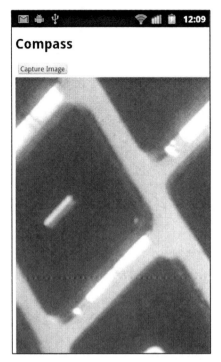

Figure 2–55. *PhoneGap Camera application after image has been captured*

You can download the complete source for this example from
`https://bitbucket.org/rohitghatol/apress-`
`phonegap/src/67848b004644/android/PhoneGap-Camera`.

You can refer to the official documentation of Camera API at
`http://docs.phonegap.com/en/1.1.0/phonegap_camera_camera.md.html#Camera`.

Chapter 3

Setting the Environment

The PhoneGap environment can be setup in the following two manners:

- Local development environment on your machine
- Cloud build environment on PhoneGap Build

The local development environment setup includes the developer setting up environments for each mobile platform that the developer wants to launch a PhoneGap application on. This chapter covers the local environment setup in detail and hopes that the audience won't require any other documentation to run a PhoneGap application on each of the platform emulators.

In order to run the PhoneGap application on a platform specific device, users need to look at platform specific documents. References to that documentation are provided in this chapter.

On the other hand, the cloud build environment called "PhoneGap Build" allows you to build PhoneGap applications without the need for a local development environment. This means that a developer will only code the PhoneGap portion of the application, which requires HTML, JavaScript, and CSS. This code will then be provided to the PhoneGap Build service. The PhoneGap Build service will build the required binaries for each platform and the developer can download these. We examine this process in more detail in this chapter.

Local Development Environment

The local development environment is much like what we did for Android in Chapter 2. In this chapter, we will see how to setup a PhoneGap environment on your development machines for the following platforms:

1. iOS
2. BlackBerry
3. Symbian
4. webOS

It is important to note that iOS can only be built on a Mac using Xcode and, for BlackBerry, the preferred OS is Windows.

Prerequisite Steps

Before we go any further, we will complete the common steps for all of the platforms beforehand. The first step is to download PhoneGap.

Download PhoneGap

You can download PhoneGap sdk from www.phonegap.com. This book employs PhoneGap version 1.1.0, which was the latest version at that time.

Once you download PhoneGap sdk and unzip it, you will see the folder structure shown in Figure 3–1.

Figure 3–1. *PhoneGap SDK directory structure*

There is a separate directory for each of the platforms that PhoneGap supports. Each directory contains library, tools, and source code for each platform to help setup the local development environment.

Setting Environment for iOS Using Xcode 4

In order to work with iOS, you will need an Intel-based computer with Mac OS X Snow Leopard (10.6).

You will also need the following in order to test your PhoneGap application on a device:

1. An Apple device like iPhone, iPad, or iPod Touch

2. iOS Developer account and certificate

Next you will need to perform the following installation steps:

1. Install Xcode and PhoneGap. Xcode installer can be downloaded from http://developer.apple.com/xcode/index.php, an Apple Developer Portal. Please note that you will need an Apple Developer account for this. Alternatively, you can purchase and download Xcode 4 from iTunes for around $5.

2. Navigate to the iOS directory where you extracted PhoneGap sdk. Run the PhoneGap installer until completion.

3. Create a new PhoneGap project. Open Xcode and create a new project. This will present the following dialog box. Select the "PhoneGap Based Application" option and select the Next button (see Figure 3–2).

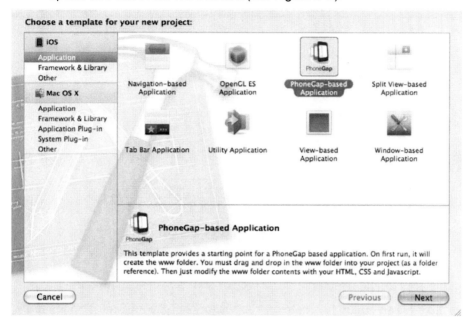

Figure 3–2. *Create a new iOS PhoneGap project*

4. On the next screen (shown in Figure 3–3), provide the product name and company identifier to the project creation wizard. Click next.

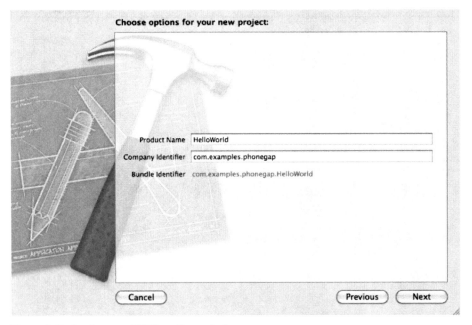

Figure 3–3. *Create a new iOS PhoneGap project*

5. Select the appropriate directory for the project and click the Create button.

Xcode provides an option to create a git repository for your project. This feature can be enabled or disabled by clicking on the source control check box shown in Figure 3–4.

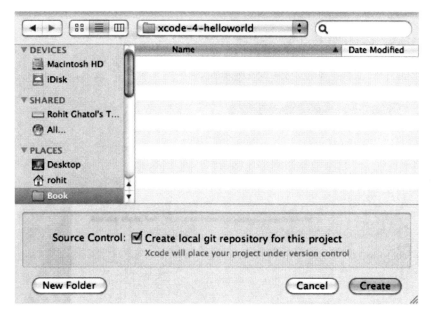

Figure 3–4. *Create a new iOS PhoneGap project*

Now you should see the HelloWorld project in Xcode.

1. Input PhoneGap's HTML and JavaScript

    Please note that there is no www folder in our project. To create a www folder,
    click on the Run button in the top left corner of Xcode. It will build the project and
    launch it in the simulator. Don't worry about the error in your simulator that says
    "index.html was not found". It is expected as we haven't put our HTML in the file
    yet.

2. Open the project in Finder (see Figure 3–5).

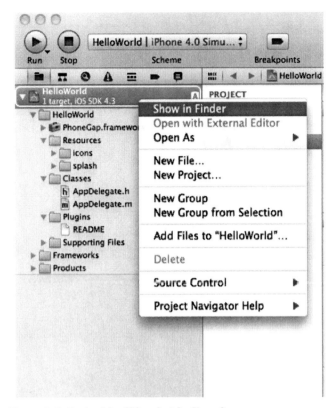

Figure 3–5. *Customizing iOS project for PhoneGap*

You will see a www folder next to the project folder. We need to copy this folder into the Xcode project.

1. Drag and drop the www folder into Xcode. Xcode will now prompt with a few options. Select Create Folder References for any added folders and click the Finish button. Now you should see the project structure shown in Figure 3–6 in Xcode.

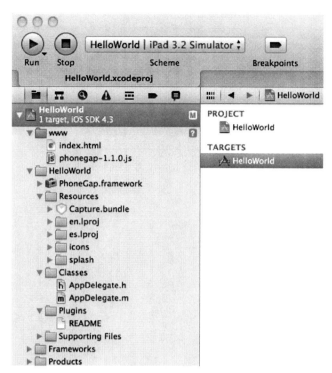

Figure 3–6. *PhoneGap WWW folder in iOS project*

2. Write PhoneGap Application

   You can write the PhoneGap application by modifying the index.html file. In order to open index.html, open the www folder and open index.html page in editor. Type your content in the index.html file. You can also specify associated Javascript and CSS files on the index.html page.

3. Deploy to the Simulator. Make sure the Simulator-version is selected as Active SDK in the top left menu.

4. Click the Run button in the Xcode project header to build the project and launch the application in the Simulator (see Figure 3–7).

Figure 3–7. *PhoneGap sample application running on iOS*

5. Deploy to Device

   You can launch the PhoneGap application on the developer device. In order to run the application on the device, open HelloWorld-info.plist and change the **BundleIdentifier**. You can get the BundleIdentifier from Apple if you have a developer license.

6. Make sure the Device-version is selected as Active SDK in the top left menu, and then click the Run button in the Xcode project header. It will build the project and launch the application in the device.

Setting Environment for BlackBerry

You will need an Intel-based computer with Windows XP (32-bit) or Windows 7 (32-bit and 64-bit) for working with BlackBerry. You will need the following software installed on your PC:

1. Java se 6 jdk 32-bit

2. Apache ant

3. BlackBerry webworks sdk v2.0+

4. Any Java IDE environment

5. Account with BlackBerry developer zone

6. Install j2sdk 6 (32 bit)

You can download the J2SDK from
`http://www.oracle.com/technetwork/java/javasebusiness/downloads/java-archive-downloads-javase6-419409.html`. Run the J2SDK installer until completion. Add the Installation_directory/J2SDK/bin in the PATH environment variable.

Next you will need to perform the following installation steps:

1. Install Apache ant

   You can download the Apache ant package from
   `http://ant.apache.org/bindownload.cgi`. The Apache ant package is a zip file.
   Extract the ant zip file and put the **Extracted_directory/apache-ant-1.8.2/bin** in
   the PATH environment variable.

2. Install BlackBerry SDK

   Download the BlackBerry Webworks SDK for Smartphone from
   `https://bdsc.webapps.blackberry.com/html5/download/sdk` site. Run the
   BlackBerry installer until completion. Typically, the BlackBerry Webwork SDK
   installer installs the files in the C:\BBWP folder. If you changed it to some other
   directory, remember the installation path, as you will need to use it in later steps.
   We recommend using the path "`c:\BBWP`", as we are using that path in the rest of
   the chapter.

3. Create New PhoneGap project

   In order to create a BlackBerry PhoneGap application, the PhoneGap framework
   provides an ant script.

   ▓ Navigate to the PhoneGap BlackBerry directory

   ▓ Run 'ant create –Dproject.path=C:\Dev\Sample' command in
     command prompt

   Note that if you are not able to run the above command, try to download the
   PhoneGap's BlackBerry callback from `https://github.com/callback/callback-blackberry/downloads` and unzip it into the PhoneGap's BlackBerry-WebWorks
   directory.

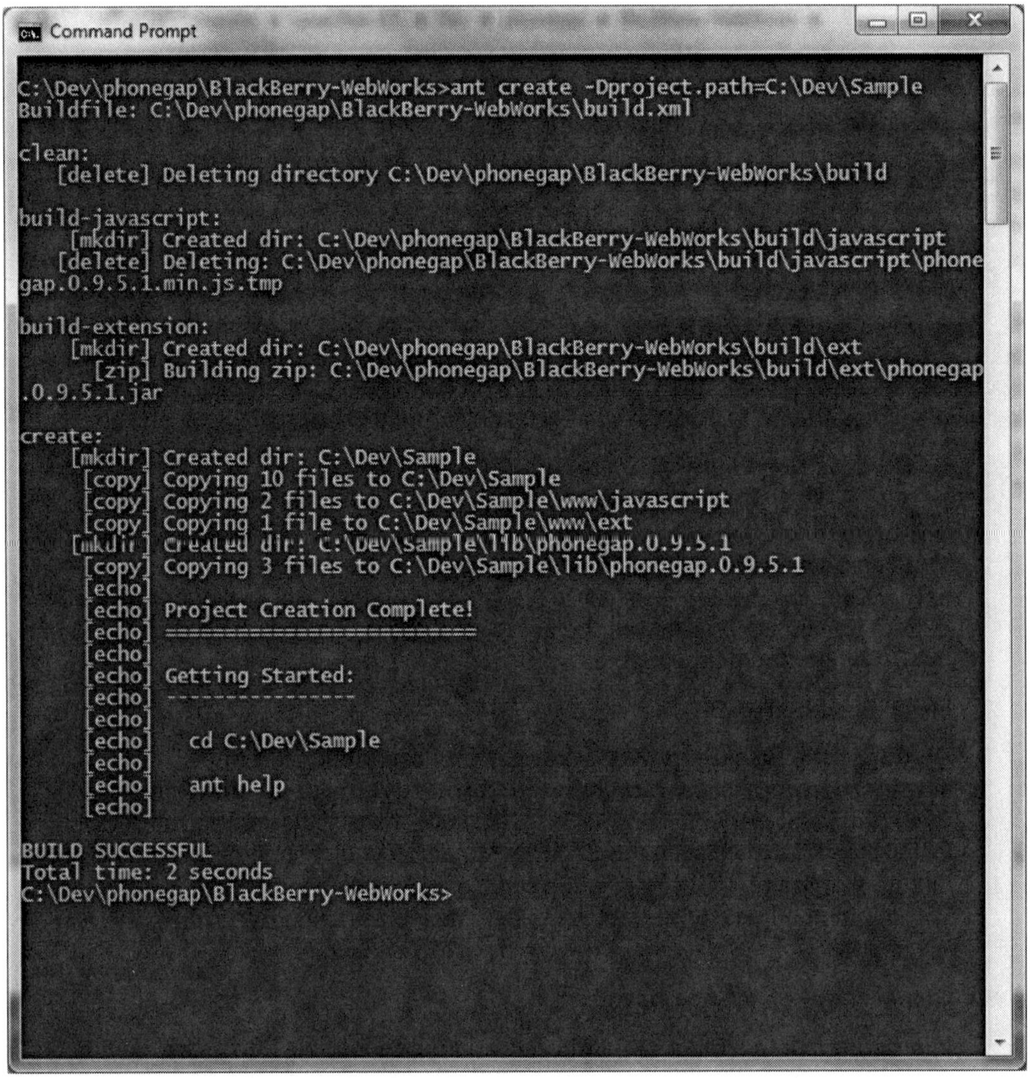

Figure 3–8. *Create BlackBerry PhoneGap project*

> This will create the project folder as depicted in Figure 3–9. Notice the www folder. This folder contains the html file and the PhoneGap JavaScript file.

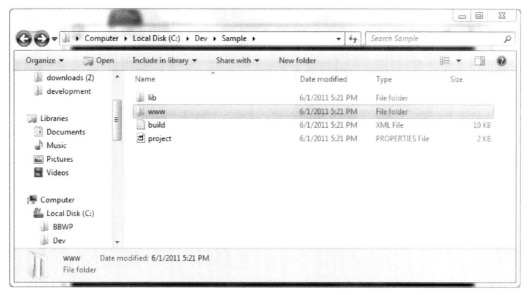

Figure 3–9. *BlackBerry PhoneGap project directory structure*

You have to change the value of the **bbwp.dir** property in the **project.properties** file to C:\\BBWP. If you changed the BlackBerry installation directory during installation, as mentioned in Step 3, make sure you are typing the same directory for the **bbwp.dir** property (see Figure 3–10).

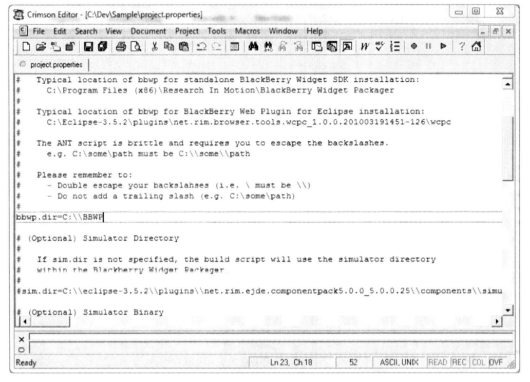

Figure 3–10. *Configuring project.properties to point to BlackBerry Works SDK directory*

4. Write a PhoneGap Application

Writing a PhoneGap application is as simple as modifying the index.html file. In order to open index.html, open the www folder and open the index.html page in your favourite editor. Include your CSS and JavaScript files in the index.html file.

5. Deploy to Simulator

The following steps are needed to deploy on a BlackBerry simulator.

- Launch the BlackBerry simulator, run the ant target as shown below. This will start the BlackBerry simulator

  ```
  C:\Dev\Sample>ant load-simulator
  ```

- Select the BlackBerry button on the simulator

- Select the Downloads folder

There you will see the PhoneGap sample application. Select it to open it (see Figure 3-11).

Figure 3–11. *Running PhoneGap App from Downloads of BlackBerry Simulator*

6. Deploy to Device

In order to deploy the PhoneGap application on a BlackBerry device, you need signing keys from RIM. You can use the following site to get your signing keys https://www.blackberry.com/SignedKeys.

Navigate to your project directory and run the following ant command in the command prompt:

```
C:\Dev\Sample>ant load-device
```

Setting Environment for Symbian

In order to work with Symbian, you will need an Intel-based computer with Windows OS. Although the official documentation of PhoneGap claims that the Symbian application can be developed on all OSs, we would recommend using Nokia Symbian s60 sdk on Windows for testing PhoneGap application on a Symbian emulator.

Next you will need to perform the following installation steps:

1. Install Cygwin

   In order to setup the environment for Symbian, we will need Cygwin installed on Windows. Download the Cygwin.exe file from `http://cygwin.com/install.html` and begin installation until completion. Please note that you must choose two packages namely zip and make packages while installing Cygwin.

2. Install the Symbian s60 sdk

   Download the Symbian s60 sdk from `http://www.forum.nokia.com/info/sw.nokia.com/id/ec866fab-4b76-49f6-b5a5-af0631419e9c/S60_All_in_One_SDKs.html` . Please note, this sdk is around 800+ mb and installation requires about 3+ gb of space. The installation will take around 30+ minutes.

3. Create a New PhoneGap project

   PhoneGap directory has a folder named Symbian in it; this directory is a template project. In order to create a new Symbian PhoneGap project, simply copy this directory and past it where you wish to create a new Symbian PhoneGap project. The contents of this directory are shown in Figure 3–12.

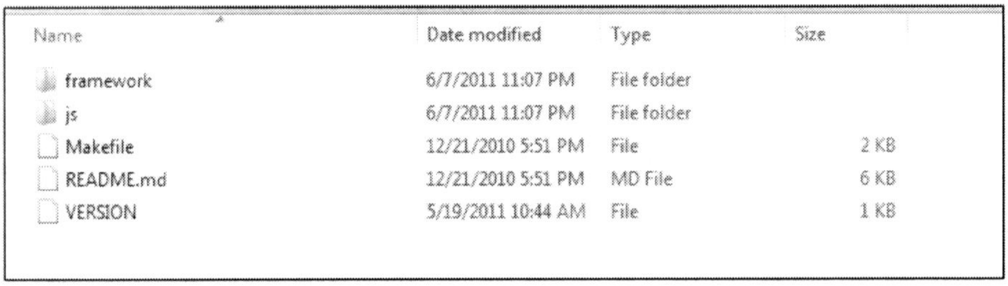

Name	Date modified	Type	Size
framework	6/7/2011 11:07 PM	File folder	
js	6/7/2011 11:07 PM	File folder	
Makefile	12/21/2010 5:51 PM	File	2 KB
README.md	12/21/2010 5:51 PM	MD File	6 KB
VERSION	5/19/2011 10:44 AM	File	1 KB

Figure 3–12. *Create new PhoneGap project*

4. Write PhoneGap Application

   Open the www folder and open the index.html in your favourite editor. Edit HTML content and include CSS and JavaScript as needed.

5. Deploy to Simulator

   Symbian PhoneGap uses makefile to build the project. On Mac or Linux machines, these can be built by simply running make in terminal. In Windows, you will need Cygwin to build. Simply run "make" in terminal/Cygwin and the Symbian project is built and the "wgz" file is created. These steps are shown in Figure 3–13.

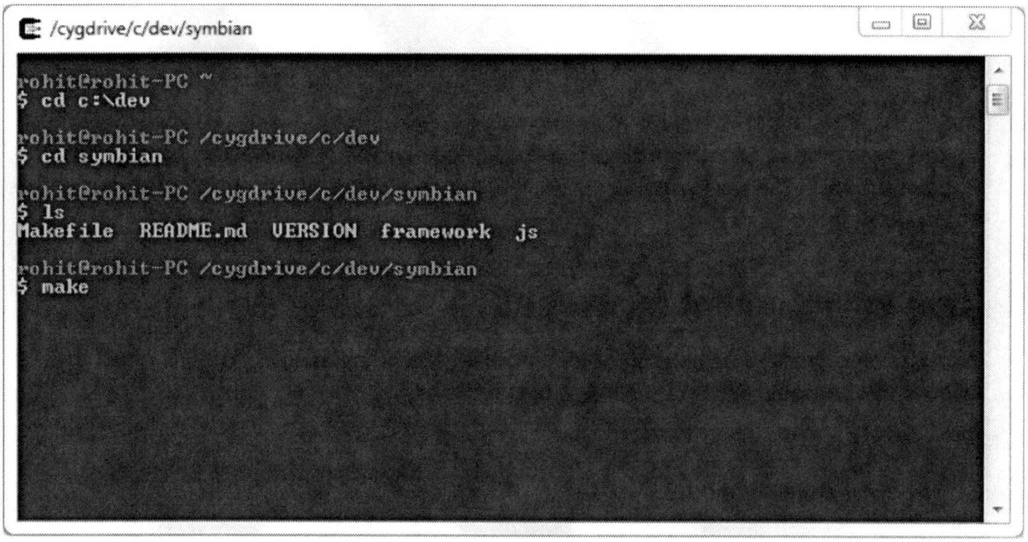

Figure 3–13. *Build Symbian project*

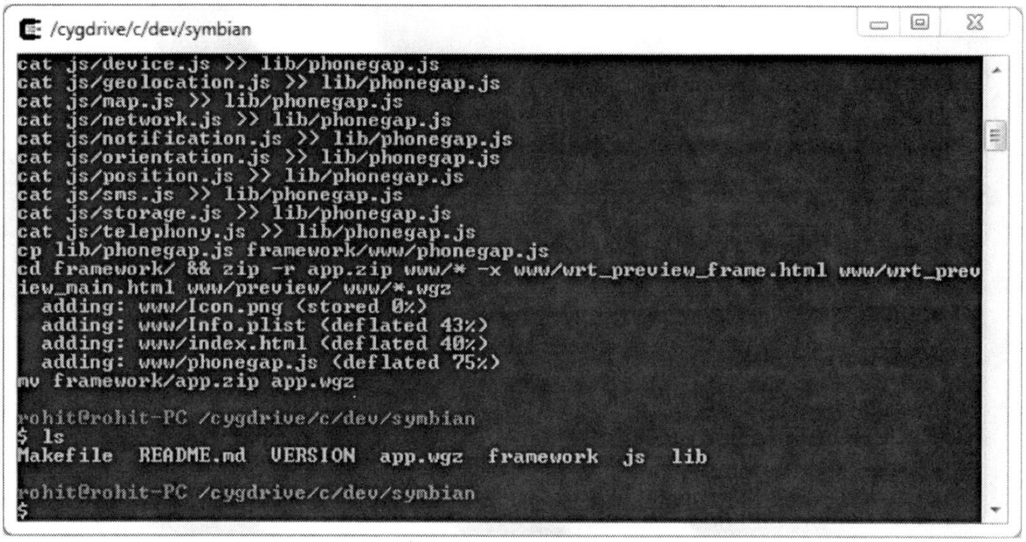

Figure 3–14. *Build Symbian project*

The app.wgz file needs to be loaded into the Symbian emulator. Use file options of the emulator to import the .wgz file. This will prompt you to install the application. Select yes to install the application.

Once the application is installed, we need to click on the middle bottom button of the Symbian emulator to see all the applications that have been installed. Launch our application from this screen. Once you launch the application, you will be prompted about the permission this application requires to run. Allow the application to use the required feature.

6. Deploy to Device

You need to use Bluetooth or email to deploy Symbian PhoneGap project to a device. Load the app.wgz into the device using Bluetooth or email and launch the application.

Setting Environment for webOS

You can develop webOS application on Windows, Mac, and Linux. You will need the following software installed on your development box:

1. Java se 6 jdk 32-bit

2. Virtual machine version 3.0 to 3.2

3. webOS sdk version 3.0.4

Next you will need to perform the following installation steps:

1. Install java se 6 jdk 32-bit

You can download the J2SDK from www.oracle.com/technetwork/java/javasebusiness/downloads/java-archive-downloads-javase6-419409.html. Run the J2SDK installer until completion. Add the Installation_directory/J2SDK/bin in PATH environment variable.

2. Install Virtual Box

Download virtual box 3.0 – 3.2 from www.virtualbox.org/wiki/Download_Old_Builds_4_0. Start installation until completion.

3. Install webOS SDK

Download webOS sdk from https://developer.palm.com/content/resources/develop/sdk_pdk_download.html. Begin installation until completion.

4. Install Cygwin for Windows only

If you are using Windows, you have to install Cygwin to build and deploy the PhoneGap application for webOS. Please look at step 1 of the Symbian installation.

5. Create new PhoneGap project

PhoneGap directory contains a directory named webOS. This is PhoneGap webOS's template project. In order to create a PhoneGap webOS project, copy this directory to your project area.

6. Write PhoneGap application

   Open www folder and open index.html in your favourite editor. Edit HTML content and include CSS and JavaScript as per your need.

7. Deploy to Simulator

   Before you deploy the project, ensure your webOS emulator is running. Run the palm-emulator from your application folder/start menu.

   Run "make" in project folder. This will create the final javaScript file. Package the project to the webOS mobile app package and install it in the webOS emulator.

8. Deploy to Device

   In order to deploy the PhoneGap project in the webOS device, you have to enable 'Developer mode' and plug it in. Run "make" in project folder in Cygwin terminal.

Cloud Build Environment Using PhoneGap Build

Until now, we saw it as tedious to build a PhoneGap application on different mobile platforms. While PhoneGap development eases the pain of cross platform mobile application development, it is still tedious for developers to build PhoneGap on each mobile platform.

To ease this pain, PhoneGap has launched PhoneGap Build. PhoneGap Build is a cloud build service. The developer submits their PhoneGap application code to PhoneGap Build, and PhoneGap Build develops the application on the following:

1. iOS

2. Android

3. Blackberry

4. webOS

5. Symbian

In this section, we will see how to set up an account on PhoneGap and build applications on it.

Registering with PhoneGap Build

1. The first step is to get an account for PhoneGap beta. Go to http://build.phonegap.com and provide your details.

2. Once you submit your details, you will get an email from PhoneGap. In this email, PhoneGap will provide the beta code. You need to provide the beta code at the signup page of PhoneGap.

Although PhoneGap Build will build applications for you, you need to own these applications. This is necessary because you will be publishing the builds you get from PhoneGap Build on Appstore, Android Market, and BlackBerry market.

Let's try to understand what it means to own your application. An application needs to be signed by certain certificates, imprinting your ownership on the application. For platforms like iOS, you need to get a developer account and get the certificate from Apple.

PhoneGap requires these certificates from you to build applications for the following:

1. iOS

2. Android

3. BlackBerry

The following sections will demonstrate how to generate these certificates and provide them to PhoneGap Build.

Registering Your Application with PhoneGap Build

The first step toward PhoneGap Build is to register your application with PhoneGap Build. There are three ways you can register with PhoneGap Build.

1. Create a new git repository on PhoneGap server and push code there.

2. Pull code from existing git repository.

3. Upload an archive of you PhoneGap app.

For simplicity, we will go with the second option of pulling the starter PhoneGap project from PhoneGap git repository (as shown in Figure 3–15).

> **NOTE:** PhoneGap Build requires access to your source code, as it's a build tool. This means you will need to share your source code with PhoneGap Build. This is like saying we want to build PhoneGap sample code provided by PhoneGap in one of their git repositories. This way we don't need to provide any source code and we will practice using sample source code.

Figure 3–15. *Pulling PhoneGap starter code from PhoneGap repository*

Now you can see your application listed under "your apps" section. Note that for every platform, there is a download icon (as shown in Figure 3–16). Clicking on the download button allows you to download the platform specific binary. All these binaries are built on PhoneGap Build server, thereby not requiring you to setup environments for iOS, Android, BlackBerry, Symbian, and webOS.

Figure 3-16. *PhoneGap Starter Project Build download screen*

Notice the orange warning for iOS build in Figure 3-16. For iOS, the build needs to be signed by a developer certificate and a provisioning profile. The provisioning profile is linked to an Apple developer account. If a developer wishes to test the application on an iOS device, the device needs to be registered with this provisioning profile.

Now, out of the above 5 platforms listed, we need to provide some kind of developer private keys to PhoneGap Build for 3 platforms. This is required in order to get platform specific builds, which can be:

1. Installed on devices.

2. Uploaded to respective app store.

These platforms are:

1. Android

2. iOS

3. BlackBerry

Setting Up Android Build Environment

Android applications are signed by self-signed keystore before they are published to the Android market. Android does not need a central authority to certify the developer's application. However, if version 1 of an application is signed with an xyz developer keystore, then the next version of the application has to be signed with the same developer keystore. Failing to do so will result in the next version of application being rejected by Android Market.

The following are the steps in order to configure the PhoneGap Build to turn out a proper Android build, which can be deployed on Android Market:

1. Create private keystore

2. Upload the private keystore to PhoneGap Build

3. Run PhoneGap Build

1. Create Private Keystore

The starting point for this is to understand the Android application publishing guidelines listed at `http://developer.android.com/guide/publishing/app-signing.html#cert`. In this section, we will walk you through the steps required to create the private keystore.

The requirement for creating a private keystore is that you need to have java jdk 1.6 and above installed on your machine. Confirm this by opening a terminal/command prompt and typing the following:

```
$> keytool
```

If this shows you some kind of help, then you are good to go. If this tells you that there is no tool named keytool, make sure your java bin directory is in path.

Now, to actually create your private keystore, follow the steps provided below:

```
$> keytool -genkey -v -keystore my-release-key.keystore
```

You will be prompted for a password. Enter a password, and write it down.

```
$>Enter keystore password: welcome
```

You will be asked to re-enter the password.

```
$>Re-enter new password: welcome
```

Next you will be asked a number of questions to record your identity in the private keystore. Please answer these questions.

$>What is your first and last name?

```
  [Unknown]:  Rohit Ghatol
$>What is the name of your organizational unit?
  [Unknown]:  Engineering
```

```
$>What is the name of your organization?
  [Unknown]:  QuickOffice
$>What is the name of your City or Locality?
  [Unknown]:  Pune
$>What is the name of your State or Province?
  [Unknown]:  Maharahstra
$>What is the two-letter country code for this unit?
  [Unknown]:  IN
```

Now you will be asked to confirm that the data you have entered so far is correct.

```
$>Is CN=Rohit Ghatol, OU=Engineering, O=QuickOffice, L=Pune, ST=Maharahstra, C=IN
correct?
  [no]:  yes
```

> **NOTE:** Tool will ask you password for creating the self-signed certification. To keep this the
> same as a previous password, just hit enter.

```
$>Generating 1,024 bit DSA key pair and self-signed certificate (SHA1withDSA) with a
validity of 90 days
        for: CN=Rohit Ghatol, OU=Engineering, O=QuickOffice, L=Pune, ST=Maharahstra,
C=IN
$>Enter key password for <mykey>
        (RETURN if same as keystore password):
 [Storing my-release-key.keystore]
```

Now your private keystore file named "my-release-key.keystore" is created and kept in
the same directory.

2. Upload the Private Keystore to PhoneGap Build

Navigate to your application on PhoneGap Build, hit the edit button, and you will see the
screen shown in Figure 3–17.

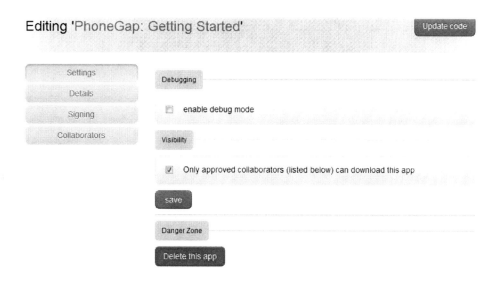

Figure 3–17. *PhoneGap Build Edit Application Screen*

Navigate to the signing section and upload the Android keystore information on that screen (as shown in Figure 3–18)

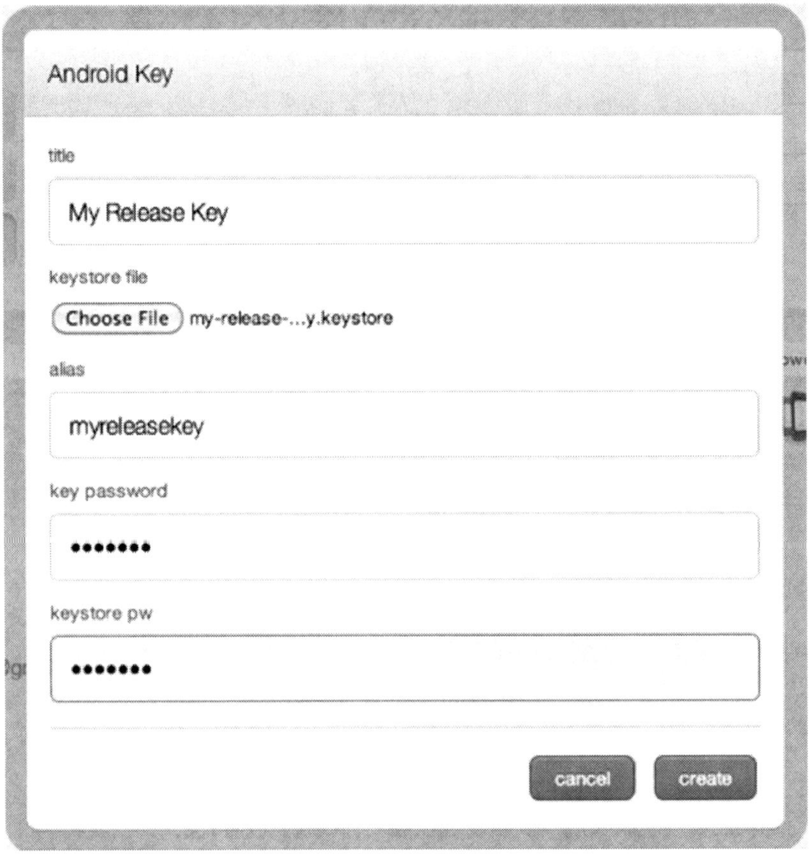

Figure 3–18. *Enter Android Release keystore details*

Here provide a title for your keystore. On PhoneGap Build, you can provide multiple keystore and choose which one you will use to build your application. The title is to recognize the keystore when you are browsing various keystores you uploaded on PhoneGap Build.

Next Upload the keystore file and provide any alias. In the password fields (both), you need to put in the password, which you used to create the keystore and private certificate (in our case this was "welcome").

3. Run PhoneGap Build

Lastly, hit create and PhoneGap Build will store this keystore on your behalf.

Now you should see a screen that looks like Figure 3–19, telling you that "my release key" keystore is used to build the Android build.

Figure 3–19. *Android keystore registered for PhoneGap app*

Setting Up iOS Build Environment

Before we start, let's note the list of the prerequisites for iOS build on PhoneGap.

1. Apple developer account either for Apple developer program
   (http://developer.apple.com/programs/ios/ 0) or Apple Enterprise Developer
   Program (http://developer.apple.com/programs/ios/enterprise/). Choose the
   one that best suits your requirements.

2. Mac Machine with Xcode to extract the developer certificate and provisioning
   profile. After this information is extracted, the developer can use PhoneGap Build
   from any Operating System.

The next steps are as follows:

1. Get the iOS keys
2. Provide the iOS keys to PhoneGap Build

1. Getting iOS Keys

The complete information about how to configure your Xcode with your iOS developer
account is described in detail on http://tiny.cc/appleprov.

Follow the above steps and ensure you are able to build and install a sample iPhone
Application on an emulator, preferably on an iOS device (already added to your
provisioning profile).

Now the next step is to export the developer certificate and provisioning profile from
Xcode and put them in PhoneGap Build.

2. Providing iOS Keys to PhoneGap Build

After setting up the iOS development environment properly, we need to extract the
developer certificate and mobile provisioning profile and upload them to PhoneGap
Build.

The first step is to extract the developer certificate from Mac's keychain access. Open keychain access and located the developer certificate and export it out. While exporting, you will be asked for a folder and a password. Record this password, as you will need the password when you upload the developer certificate on PhoneGap Build site.

After this, we will proceed to extract the provisioning profile. The provisioning profile is inside Xcode. Open Xcode and go to window->organizer and launch it. From here, export the "team provisioning profile". Make a note that the provisioning profile is where you tell Apple that you have registered your iPhone/iPod/iPad as a developer device to test your application. PhoneGap Build needs this profile to sign your application (ipa).

We have now extracted both the developer certificate and the mobile provisioning profile in a directory called ios-Keys.

Figure 3–20. *Directory containing Developer Certificate and Provisioning Profile*

We need to upload these keys on PhoneGap Build. Again visit the edit screen of your application, navigate to the signing section, and, for iOS, click on add keys (below select a key drop down).

Figure 3–21. *iOS Add Key Screen*

This will open the iOS Certificate and provisioning profile Pair screen. Upload the needed keys and the same password that you used to export the developer certificate.

Figure 3–22. *Provide PhoneGap Build with Developer Certificate and Provisioning Profile*

Once you fire up PhoneGap Build, you should notice the orange warning on iOS ipa build has gone and the build has turned green. Clicking on the ipa button will bring down the iOS ipa for you.

Setting Up BlackBerry Build Environment

The main actions required for setting up your BlackBerry build environment are as follows:

1. Get the BlackBerry keys

2. Provide the BlackBerry keys to PhoneGap Build

1. Getting BlackBerry Keys

PhoneGap Build provides out of the box support for building BlackBerry applications, which can be installed on your devices. However, in order to upload these applications for distribution, you need to have keys provided from rim. In order to gain these keys, you need to register with rim using this web site - https://www.blackberry.com/SignedKeys/.

Once you register, you will get an email from rim mentioning the steps to install these keys with your BlackBerry development environment. We cannot go into details of the instructions due to legal prohibitions with respect to sharing those instructions.

2. Providing BlackBerry Keys to PhoneGap Build

The next step is to extract the keys from the BlackBerry development environment. The BlackBerry development environment can be setup using either eclipse or standalone BlackBerry web works. The BlackBerry keys are located in the sdk directory of the BlackBerry.

The first task is to locate the sdk directory.

If the BlackBerry development environment was installed using eclipse, you should find the BlackBerry sdk directory at the `<<eclipse location>>\plugins\ net.rim.ejde.componentpackX.X.X_X.X.X.X \components`. An example of that is d:\worksoft\eclipse-helios\ net.rim.ejde.componentpack5.0.0_5.0.0.25\components.

If the BlackBerry development environment was installed using BlackBerry Widget/WebWorks Packager Standalone SDK (like how we showed in the earlier part of this chapter), the installation directory is the sdk directory. This was shown as "c:\BBWP" directory in the earlier part of this chapter.

The code signing files/keys are located in the sdk directory.

`<<webworks_sdk_dir>\bin\sigtool.csk`

`<<webworks_sdk_dir>\bin\sigtool.db`

Now that we have gained access to the BlackBerry keys, proceed to your app on PhoneGap Build, click edit, and go to the signing section. From the drop down menu, choose "add a key" option for BlackBerry.

Figure 3–23. *iOS Add Key Screen*

Upload the BlackBerry keys in the dialog box shown below. Use the same password you created while following instructions, which you got in the email from BlackBerry. This provides PhoneGap Build with all information needed to build BlackBerry applications, which can be distributed on BlackBerry distribution channels.

Figure 3–24. *BlackBerry Keys files upload*

Launching PhoneGap Build

PhoneGap Builds can be fired in two ways:

1. Manually hitting "rebuild all" on PhoneGap Build as depicted below. This will queue up builds on the PhoneGap Build server.

2. The second way is to use PhoneGap Build restful api to create app, update code, and fire builds. You can make PhoneGap Builds part of your cit builds (from bamboo or jerkins or any cit systems). Make the cit build script call PhoneGap Build restful api. The details of the PhoneGap Build api can be found at `https://build.phonegap.com/docs/api`.

Conclusion

As cloud is increasingly becoming popular in hosting web services and applications, many companies are looking at cloud-based saas services to help in their development cycle.

Companies, like pivotal tracker, are used for agile planning, bitcode to host source code, and there are even online cit builds (e.g., jira studio). The advent of PhoneGap Build comes as no surprise.

Saving the infrastructure cost by not buying additional Windows and Mac machines just for cit makes sense. Keeping cost low and hiring infrastructure (just like we pay for electricity as utility bills) is becoming the trend of the day.

PhoneGap Build fits into the need of small to mid-size companies who want to stay out of the infrastructure burden and use cloud-based saas service for building PhoneGap applications.

Chapter 4

Using PhoneGap with jQuery Mobile

While PhoneGap provides a platform to allow JavaScript apps to access native phone features, there are many other things that contribute to a mobile HTML app.

One of the most important parts of the mobile HTML application is the UI. You could write the entire UI by hand using HTML, JavaScript, and CSS. However, any web developer will tell you that there are many issues with this approach, including the following:

1. Not all browsers are same; you need a cross-browser framework to be successful. Even if most mobile browser are webkit based, its best to use a framework that abstracts the browser differences from a developer.

2. If you were coding by hand, most of your code would be of drawing the UI, modifying the DOM, and making Ajax calls. A framework that lets you write less and do more would help you actually focus on the business logic.

3. Creating an aesthetically good-looking HTML UI requires designer skills. At the same time, most mobile clients have predefined themes or schemas. It would help a developer if a framework provides good-looking UI out of the box. That way, the developer could focus on the business logic.

Having said that, one of the easiest frameworks to use with PhoneGap to write your UI is jQueryMobile. First of all, jQueryMobile is built on top of the very popular jQuery. jQuery is known to be a JavaScript library that increases developer productivity and helps developers with cross-browser compatibility. At the same time, there are many free plug-ins available with jQuery to do a lot of things.

jQueryMobile is a UI framework built for a mobile UI. It has a declarative UI, which means you don't have to code your UI in JavaScript, but can declare it in HTML. jQueryMobile also provides an excellent looking UI out of the box.

All this makes jQueryMobile the most easy to use JavaScript UI framework and the most appropriate framework for a mobile UI of moderate complexity.

Having said that, jQueryMobile provides the same UI for smartphones and tablets. If your needs are different, and you need different layouts for smartphone and tablets, you should look at Chapter 5.

Getting Acquainted with jQuery

jQuery is one of the best JavaScript libraries that help you do Ajax calls, search the HTML DOM for a particular element, and modify the DOM. It also has its own plug-in framework. The best thing is its cross-browser framework, taking the headache out of browser differences. You can refer to the following jQuery tutorial: www.w3schools.com/jquery/default.asp.

jQuery Initialization

jQuery initialization is a two-step process.

1. Include jQuery JavaScript in your HTML page.

2. Declare a callback, which will be called by jQuery when jQuery's library is loaded.

This is necessary as an HTML page may include many files like CSS, JavaScript, and images. The browser would download all these resources and start executing all the JavaScript blocks. If you start to use jQuery API calls before proper initialization, you will get errors. Therefore, you declare a callback, which is an entry point for our application, and jQuery will call this callback and bootstrap the application.

Typically, when a developer is not using jQuery, he would write the code in the following manner:

```
window.onload = function(){
    alert("Page Loaded");
}
```

When using jQuery the same code would look like follows:

```
<html>
    <head>
        //Step 1 - include jquery library
        <script type="text/javascript" src="jquery.js">

        </script>
        <script type="text/javascript">
                //Step 2 - Declare a callback for jquery to call when it loads
                $(document).ready(function() {
                    //Place to bootstrap your application
                    alert("jquery loaded");
                });
        </script>
    </head>
```

```
    <body>
        <h1>
            jQuery Demo
        </h1>
    </body>
</html>
```

When you run this code in any browser, you will see an alert popping up when the page loads. The alert would say "jQuery loaded."

jQuery Selectors

Now that you have seen how to initialize jQuery and register an onload() method, let's move on to how to find HTML DOM elements.

Typically, a developer would use following code to get a div with id "placeholder."

```
document.getElementById("placeholder").innerhtml = "hello world";
```

In jQuery, you would write the above function as

```
$("#placeholder").html("hello world");
```

jQuery provides a number of ways to locate an HTML element. One example is $("#placeholder"), which returns a jQuery element wrapping an element that has id "placeholder." Once you get this jQuery wrapper, you can call jQuery functions to manipulate the DOM. In the above case, you are changing its HTML content to "hello world."

Let's review some other useful examples of selectors with reference to a code example:

```
<html>

    <body>
        <h1 class="title">
            JQUERY SELECTOR Tutorial
        </h1>
        <p>
            simple paragraph
        </p>
        <p class="title">
            Paragraph with class title
        </p>
        <p>
            another paragraph
        </p>
        <ul id="selector">
            <li>
                Element based - $("p")
            </li>
            <li>
                Id based - $("#selector")
            </li>
            <li>
                CSS Class based - $(".title")
            </li>
```

```
            <li>
                Element + Class based - $("p.title");
            </li>
            <li>
                Element+ID+Position - $("ui#selectorli:first)
            </li>
        </ul>
    </body>

</html>
```

Element-Based Selector

$("p") selects all the paragraphs:

```
<p>simple paragraph</p>
<p class="title">Paragraph with class title</p>
<p>another paragraph</p>
```

ID-Based Selector

$("#selector") selects the element with id "selector." Remember, # is added before the id that you want to search. The following element is selected for this selector:

```
<ul id="selector">
```

CSS-Based Selector

$(".title") selects the element with class "title." Remember, a "." is called before a class name to search elements with that class. The following elements are selected for this selector:

```
<h1 class="title">JQUERY SELECTOR Tutorial</h1>

<p class="title">Paragraph with class title</p>
```

Combination of Selectors

The following are some examples of how to mix and match selectors to pin point a particular element:

$("p.title") selects a paragraph element with class "title," which selects the following element:

```
<p class="title">Paragraph with class title</p>
```

$("ul#selector li:first") selects the first li element from a UI with an id "selector." This selects the following element:

```
<li>Element based                - $("p")</li>
```

A complete list of jQuery selectors can be found at the following link: www.w3schools.com/jquery/jquery_ref_selectors.asp.

jQuery DOM Manipulation

First, let's look at how to retrieve values from HTML. You can retrieve the value of an element or the inner HTML.

If you execute javascript `$(ul#selector).html()`, you will get the following text:

```
<li>Element based                - $("p")</li>
<li>Id based                      - $("#selector")</li>
<li>CSS Class based              - $(".title")</li>
<li>Element + Class based      - $("p.title");</li>
<li>Element+ID+Position        - $("ui#selector li:first)</li>
```

The following example shows how to extract value from a jQuery selector result. First, the example shows what happens when you assume the jQuery selector returns only one value. That is, assume $("p") returned only one paragraph. You will then try to get the value of the paragraph using the html() function. Note that $("p") returns a jQuery Selector and the html() function belongs to the jQuery Selector. In this case, jQuery will run the html() method on the first element found by $("p") selector. This means when you say $("p").HTML(), $("p") will locate the first paragraph as below and give the value "simple paragraph."

```
<p>simple paragraph</p>
```

Now you know that $("p") should give you many values as there are many paragraphs in the previous example, as shown here:

```
<p>simple paragraph</p>
<p class="title">Paragraph with class title</p>
<p>another paragraph</p>
```

In order to iterate over any list, jQuery provides a method: "each()." In order to use "each()," you need to invoke it on a jQuery selector like in our case $("p"). The "each()" method takes two arguments: the first being the index (position in the iteration) and the second being the actual entry at that position.

Therefore, when you say $("p").each(function(index,element){}), the element is the actual jQuery selector on each paragraph.

```
<script type="text/javascript">
$(document).ready(function() {
    alert("Simple extraction = " + $("p").html());

    $("p").each(function(index, element) {
        alert(index + " - '" + $(element).html() + "'");
    });

});
</script>
```

Now, let's move on to modifying HTML DOM. We will quickly introduce the simplest method to manipulate the DOM.

When you execute the following JavaScript, it will modify the content of paragraph to "Changed to 123."

```
<script type="text/javascript">
$(document).ready(function() {
    $("p.title").html("Changed to 123");
});
</script>
```

A complete list of jQuery HTML operations can be found at www.w3schools.com/jquery/jquery_ref_html.asp.

jQuery Ajax Calls

jQuery provides a number of useful methods for doing Ajax calls.

The following is an example of doing an Ajax GET call to a URL. This is a classic example of "write less and do more." The following code does an Ajax GET call to service/employee/details.txt and puts the contents in div with id "details":

```
$.get("service/employee/details.txt", function (result) {
    $("div#details").html(result);
});
```

The following is an example of doing an Ajax POST call to a URL, posting the data {name:employeeName}:

```
$.post("service/employee/details", {
    name: employeeName
}, function (result) {
    alert("Post successful");
});
```

A complete list of jQuery HTML operations can be found at www.w3schools.com/jquery/jquery_ref_ajax.asp.

Getting Acquainted with jQueryMobile

jQueryMobile takes the jQuery concept of "write less, do more" to the next level by providing a common UI platform to develop mobile applications across many popular mobile platforms.

jQueryMobile is built on very popular and robust jQuery and jQuery UI framework. jQueryMobile gives out-of–box, touch-ready mobile widgets such as list view, a header with a back button, navigation animation, and many more things. These widgets have a professional and polished look and feel, making it easier to develop ready-to-ship finished apps.

jQueryMobile's homepage is http://jquerymobile.com/.

Moreover, jQueryMobile offers five out-of-box themes for you to choose from. The following is an example that shows how the buttons look in different themes. Overall, we have five themes—Theme a, Theme b, Theme c, Theme d, and Theme d—as shown in Figure 4–1.

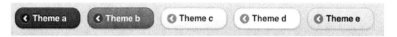

Figure 4–1. *jQueryMobile themes*

Also, jQueryMobile provides grade support for the platforms listed in Table 4-1.

Table 4–1. *jQueryMobile Supported Platform*

OS Platform	Platform Version
iOS	Version 3.1.3 onwards
Android	Version 1.5 onwards
Symbian	S60 Version 5 onwards
BlackBerry OS	Version 5.0 onwards
WebOS	Version 1.4.1 onwards
Windows 7	Version 7.0 onwards
Samsung Bada	Version 1.0 onwards
MeeGo	Version 1.1 onwards

Including jQueryMobile in Mobile App

Download jquery.mobile-1.0rc2.zip from http://jquerymobile.com/download/ and unzip it. Once you unzip it, you will see a folder structure as shown in Figure 4–2. It contains two pairs of jQueryMobile JavaScript and a CSS file. As the name suggests, you would use the .min JavaScript and CSS file in production as they are minified JavaScript and CSS files.

Along with that, you will need to include the images folder in your mobile app.

images
jquery.mobile-1.0rc2.css
jquery.mobile-1.0rc2.js
jquery.mobile-1.0rc2.min.css
jquery.mobile-1.0rc2.min.js

Figure 4–2. *jQueryMobile folder structure*

The following is the HTML template for jQueryMobile examples:

```
<!DOCTYPE HTML>
<html>

    <head>
        <title>
            jQuery Mobile Demo
        </title>
        <link rel="stylesheet" type="text/css" href="jquery.mobile-1.0rc2.min.css"/>
        <script type="text/javascript" src="jquery-1.6.4.min.js"></script>
        <script type="text/javascript" src="jquery.mobile-1.0rc2.min.js"></script>
    </head>

    <body>
        …
    </body>

</html>
```

jQueryMobile Declarative UI

Declarative UI building is the best part of jQueryMobile. You don't need to write complex JavaScript code to build the UI. UI building is just like adding normal HTML elements with some jQueryMobile specific attributes and their values.

Pages and Dialogs

You saw the HTML template in the previous section. Now, you will add jQueryMobile layouts and widgets to it.

You can declare the pages using data-role attributes in the div element inside the body tag. Thus, inside a body tag, you can declare many pages by just declaring the following:

```
<div data-role="page"></div>
```

In the same manner, you can declare the components of a page by declaring data-role as "header," "content," and "footer." The following is an example of a page in jQueryMobile:

```
<div data-role="page">

    <div data-role="header"></div>

    <div data-role="content"></div>

    <div data-role="footer"></div>

</div>
```

The complete example of a page is shown here.

```
<!DOCTYPE HTML>
<html>
    <head>
        <title>jQuery Mobile Demo</title>
        <link href="jquery.mobile-1.0rc2.min.css" rel="stylesheet" type="text/css"/>
        <script src="jquery-1.6.4.min.js"></script>
        <script src="jquery.mobile-1.0rc2.min.js"></script>
    </head>

    <body>
        <!-- Page Start-->
        <div data-role="page">
            <!-- Page Header Start -->
            <div data-role="header">
                <h1>Page Title</h1>
            </div>
            <!-- Page Header End -->

            <!-- Page Body Start -->
            <div data-role="content">
                <p>
                    Page content goes here.
                </p>
            </div>
            <!-- Page Body End -->

            <!-- Page Footer Start -->
            <div data-role="footer">
                <h4>
                    Page Footer
                </h4>
            </div>
            <!-- Page Footer End -->
        </div>
        <!-- Page End -->
    </body>

</html>
```

This html when run in browser shows a screen as depicted in Figure 4–3.

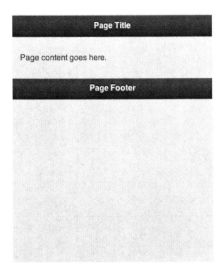

Figure 4–3. *jQueryMobile pages*

Now that you have seen the concept of a page in jQueryMobile, let's go over a scenario in which you have multiple pages or dialog box.

A typical application has multiple pages and dialog boxes. The best thing about jQueryMobile is that you can define all these different pages and dialog boxes within the same HTML page.

Here is how to do it. Define multiple divs in an HTML page and give them the following:

1. An attribute named data-role set to page. This looks like this: data-role="page"

2. An attribute named "id" to identify them in code

You use links and buttons to navigate to these pages and dialog boxes. The simplest way to do that is to do the following:

1. Define a link with href as #+<<id of the page/dialog box>>

2. Give that link a data-role="button"

Note that a declaration of a page and dialog is the same. In fact, there is nothing called as a dialog, but you can load a page in a pop-up as a dialog.

The following is an example of a link to a page. When this link is clicked, the transition to the page with id "page2" takes place. Also note that because the link is given the data-role="button", it looks like a button.

```
<a data-role="button" href="#page2">Page Navigation</a>
```

Opening a page as a dialog is quite similar to navigating to page. You only need to add two more attributes to the link:

1. data-rel="dialog"

    2. data-transition="pop" (this is the animation effect)

```
<a data-role="button" href="#dialog1"  data-rel="dialog" data-transition="pop">Open
Dialog </a>
```

Here is the complete example for you to try. Figure 4–4 shows the "main" page, Figure 4–5 shows the "page2" page, Figure 4–6 shows "dialog1" page, which acts as a dialogbox:

```
<!DOCTYPE html>
<html>

    <head>
        <title>jQuery Mobile Demo</title>
        <link href="jquery.mobile-1.0rc2.min.css" rel="stylesheet" type="text/css"/>
        <script src="jquery-1.6.4.min.js"></script>
        <script src="jquery.mobile-1.0rc2.min.js"></script>
    </head>

    <body>
        <!-- Main Page-->
        <div data-role="page" id="main">
            <div data-role="header">
                <h1>
                    Main Page
                </h1>
            </div>
            <div data-role="content">
                <h1>
                    Page Nav and Dialog Example
                </h1>
                <a data-role="button" href="#page2">Page Navigation</a>
                <a data-role="button" href="#dialog1" data-rel="dialog" data-
transition="pop">Open Dialog </a>
            </div>
            <div data-role="footer">
                <h4>
                    Main Page Footer
                </h4>
            </div>
        </div>
        <!-- First Page End -->
        <!-- Second Page-->
        <div data-role="page" id="page2" data-add-back-btn="true">
            <div data-role="header">
                <h1>
                    Second Page
                </h1>
            </div>
            <div data-role="content">
                <h1>
                    Second Page
                </h1>
            </div>
            <div data-role="footer">
                <h4>
                    Click back to go back to main page
                </h4>
            </div>
```

```
        </div>
        <!-- Second Page End -->
        <!-- Dialog -->
        <div data-role="page" id="dialog1">
            <div data-role="header">
                <h1>
                    Dialog Title
                </h1>
            </div>
            <div data-role="content">
                v
                <h1>
                    Dialog body
                </h1>
            </div>
            <div data-role="footer">
                <h4>
                    Click close button to go back to main page
                </h4>
            </div>
        </div>
        <!-- Dialog End -->
    </body>

</html>
```

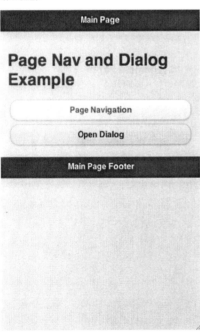

Figure 4–4. *jQueryMobile page navigation*

Figure 4–5. *jQueryMobile page navigation*

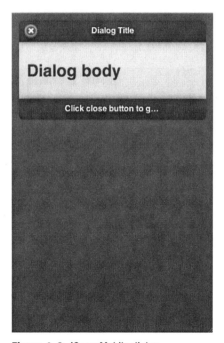

Figure 4–6. *jQueryMobile dialog*

Toolbars and Buttons

In jQueryMobile, there are two types of toolbars:

1. Header bar

2. Footer bar

In general, creating a toolbar is as simple as declaring some buttons in the header or footer bar. This is depicted in Figure 4–7.

```html
<!DOCTYPE html>
<html>

    <head>
        <title>jQuery Mobile Demo</title>
        <link href="jquery.mobile-1.0rc2.min.css" rel="stylesheet" type="text/css"/>
        <script src="jquery-1.6.4.min.js"></script>
        <script src="jquery.mobile-1.0rc2.min.js"></script>
    </head>

    <body>
        <!-- Main Page-->
        <div data-role="page" id="main">
            <div data-role="header" data-position="inline">
                <a href="index.html" data-icon="delete">Cancel</a>
                <h1>
                    Edit Contact
                </h1>
                <a href="index.html" data-icon="check">Save</a>
            </div>
            <div data-role="content">
                <h1>
                    Header Footer Toolbar Example
                </h1>
            </div>
            <div data-role="footer" class="ui-bar">
                <a href="index.html" data-role="button" data- icon="delete">Remove</a>
                <a href="index.html" data-role="button" data-icon="plus">Add</a>
                <a href="index.html" data-role="button" data-icon="arrow-
u">Up</a>
                <a href="index.html" data-role="button" data-icon="arrow-
d">Down</a>
            </div>
        </div>
        <!-- First Page End -->
    </body>

</html>
```

Figure 4–7. *jQueryMobile toolbar and buttons*

Form Elements

Form elements in jQueryMobile are typical HTML form elements—they only look different. This means you can use your HTML JavaScript skills to render good-looking, polished jQueryMobile widgets and use traditional event handling techniques to quickly write your mobile web apps.

Let's see a couple of examples of the form elements. In the first example, we are using an input text, a text area, and a search box. For all of these, a label is assigned. They are wrapped in a fieldset to form a group of the label and the associated widget. See Figure 4–8 to see how form elements look in jQueryMobile.

```
<!DOCTYPE html>
<html>

    <head>
        <title>jQuery Mobile Demo</title>
        <link href="jquery.mobile-1.0rc2.min.css" rel="stylesheet" type="text/css"/>
        <script src="jquery-1.6.4.min.js"></script>
        <script src="jquery.mobile-1.0rc2.min.js"></script>
    </head>

    <body>
        <!-- Main Page-->
        <div data-role="page" id="main">
            <div data-role="header" data-position="inline">
                <a href="index.html" data-icon="delete">Cancel</a>
                <h1>
                    Edit Contact
```

```
                    </h1>
                    <a href="index.html" data-icon="check">Save</a>
            </div>
            <div data-role="content">
                <form action="#" method="get">
                    <h2>
                        Simple Form Elements
                    </h2>
                    <div data-role="fieldcontain">
                        <label for="name">
                            Text Input:
                        </label>
                        <input type="text" name="name" id="name" value="" />
                    </div>
                    <div data-role="fieldcontain">
                        <label for="textarea">
                            Textarea:
                        </label>
                        <textarea cols="40" rows="8" name="textarea" id="textarea">
                        </textarea>
                    </div>
                    <div data-role="fieldcontain">
                        <label for="search">
                            Search Input:
                        </label>
                        <input type="search" name="password" id="search" value="" />
                    </div>
                </form>
            </div>
            <div data-role="footer" class="ui-bar">
                <a href="index.html" data-role="button" data- icon="delete">Remove</a>
                <a href="index.html" data-role="button" data-icon="plus">Add</a>
                <a href="index.html" data-role="button" data-icon="arrow-
u">Up</a>
                <a href="index.html" data-role="button" data-icon="arrow-
d">Down</a>
            </div>
        </div>
        <!-- First Page End -->
    </body>

</html>
```

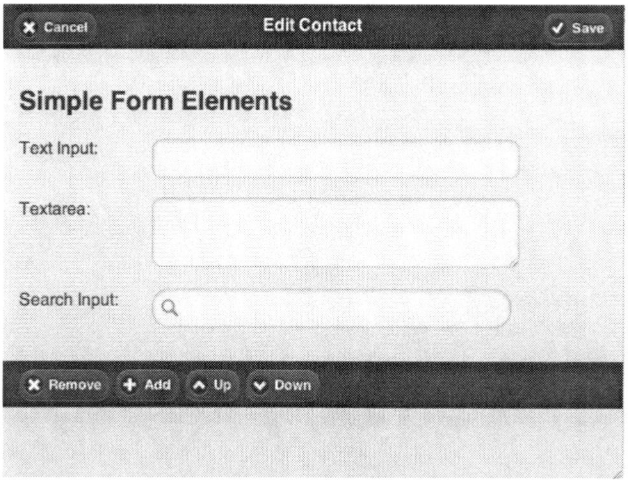

Figure 4–8. *jQueryMobile form elements*

In the second example, the HTML select is wrapped in a good-looking on/off switch. Note that when you will be fetching the value from it programmatically, you will use it as an HTML select box. You have also wrapped a textbox with a slider. The value of the slider goes in the textbox. This is as good as a user filling in a number in the textbox, but with jQueryMobile, the user can use the slider to select a value in the given range. See Figure 4–9 to see how this HTML renders.

```
<!DOCTYPE html>
<html>

    <head>
        <title>jQuery Mobile Demo</title>
        <link href="jquery.mobile-1.0rc2.min.css" rel="stylesheet" type="text/css"/>
        <script src="jquery-1.6.4.min.js"></script>
        <script src="jquery.mobile-1.0rc2.min.js"></script>
    </head>

    <body>
        <!-- Main Page-->
        <div data-role="page" id="main">
            <div data-role="header" data-position="inline">
                <a href="index.html" data-icon="delete">Cancel</a>
                <h1>
                    Edit Contact
                </h1>
                <a href="index.html" data-icon="check">Save</a>
            </div>
            <div data-role="content">
                <form action="#" method="get">
                    <h2>
                        Simple Form Elements
                    </h2>
                    <div data-role="fieldcontain">
                        <label for="onoff-slider">
                            On/Off Switch:
```

```
                                          </label>
                                          <select name="onoff-slider" id="onoff-slider" data-
role="slider">
                                                <option value="off">
                                                    Off
                                                </option>
                                                <option value="on">
                                                    On
                                                </option>
                                          </select>
                                      </div>
                                      <div data-role="fieldcontain">
                                          <label for="range-slider">
                                              Range Slider:
                                          </label>
                                          <input type="range" name="range-slider" id="range-slider"
value="0" min="10"
                                          max="100" />
                                      </div>
                                  </form>
                          </div>
                          <div data-role="footer" class="ui-bar">
                              <a href="index.html" data-role="button" data- icon="delete">Remove</a>
                              <a href="index.html" data-role="button" data-icon="plus">Add</a>
                              <a href="index.html" data-role="button" data-icon="arrow-
u">Up</a>
                              <a href="index.html" data-role="button" data-icon="arrow-
d">Down</a>
                          </div>
                  </div>
                  <!-- First Page End -->
              </body>

</html>
```

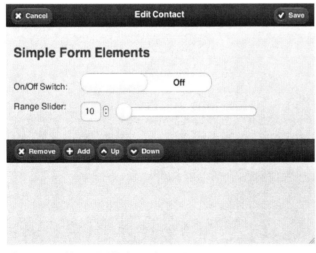

Figure 4–9. *jQueryMobile form elements*

In the following example, you can see how easy it is to wrap the HTML radio box and check box into good-looking widgets. This is done using data-role "controlgroup" and data-type as "horizontal." See Figure 4–10 to see what Single and Multi Select form elements look like:

```
<!DOCTYPE html>
<html>

    <head>
        <title>jQuery Mobile Demo</title>
        <link href="jquery.mobile-1.0rc2.min.css" rel="stylesheet" type="text/css"/>
        <script src="jquery-1.6.4.min.js"></script>
        <script src="jquery.mobile-1.0rc2.min.js"></script>
    </head>

    <body>
        <!-- Main Page-->
        <div data-role="page" id="main">
            <div data-role="header" data-position="inline">
                <a href="index.html" data-icon="delete">Cancel</a>
                <h1>
                    Edit Contact
                </h1>
                <a href="index.html" data-icon="check">Save</a>
            </div>
            <div data-role="content">
                <form action="#" method="get">
                    <h2>
                        Single and MultiSelect Form Elements
                    </h2>
                    <div data-role="fieldcontain">
                        <fieldset data-role="controlgroup">
                            <legend>
                                Choose a base:
                            </legend>
                            <input type="radio" name="radio-choice-1" id="radio-choice-
1" value="choice-1"

                            checked="checked" />
                            <label for="radio-choice-1">
                                Thin Crust
                            </label>
                            <input type="radio" name="radio-choice-1" id="radio-choice-
2" value="choice-2"

                            />
                            <label for="radio-choice-2">
                                Double Cheese Burst
                            </label>
                            <input type="radio" name="radio-choice-1" id="radio-choice-
3" value="choice-3"

                            />
                            <label for="radio-choice-3">
                                Class Hand Tossed
                            </label>
                        </fieldset>
                    </div>
                    <div data-role="fieldcontain">
                        <fieldset data-role="controlgroup">
```

```
                            <legend>
                                Choose Pizza toppings
                            </legend>
                            <input type="checkbox" name="checkbox-
1a" id="checkbox-1a" class="custom" />
                            <label for="checkbox-
1a">

                                Jalepeno
                            </label>
                            <input type="checkbox" name="checkbox-
2a" id="checkbox-2a" class="custom" />
                            <label for="checkbox-2a">
                                Olives
                            </label>
                            <input type="checkbox" name="checkbox-
3a" id="checkbox-3a" class="custom" />
                            <label for="checkbox-3a">
                                Cheese
                            </label>
                            <input type="checkbox" name="checkbox-
4a" id="checkbox-4a" class="custom" />
                            <label for="checkbox-
4a">

                                Capsicum
                            </label>
                        </fieldset>
                    </div>
                    <div data-role="fieldcontain">
                        <fieldset data-role="controlgroup" data- type="horizontal">
                            <legend>
                                Non Veg topping:
                            </legend>
                            <input type="checkbox" name="checkbox-6" id="checkbox-6"
class="custom"

                            />
                            <label for="checkbox-
6">

                                Pepperoni
                            </label>
                            <input type="checkbox" name="checkbox-7" id="checkbox-7"
class="custom"

                            />
                            <label for="checkbox-7">
                                Ham
                            </label>
                            <input type="checkbox" name="checkbox-8" id="checkbox-8"
class="custom"

                            />
                            <label for="checkbox-8">
                                Turkey
                            </label>
                        </fieldset>
                    </div>
                    <div data-role="fieldcontain">
                        <fieldset data-role="controlgroup" data- type="horizontal">
                            <legend>
                                Payment Type:
```

```html
                            </legend>
                            <input type="radio" name="radio-choice-b" id="radio-choice-
c" value="on"
                            checked="checked" />
                            <label for="radio-choice-c">
                                Cash
                            </label>
                            <input type="radio" name="radio-choice-b" id="radio-choice-
d" value="off"
                            />
                            <label for="radio-choice-d">
                                Coupons
                            </label>
                            <input type="radio" name="radio-choice-b" id="radio-choice-
e" value="other"
                            />
                            <label for="radio-choice-e">
                                Credit Card
                            </label>
                        </fieldset>
                    </div>
                </form>
            </div>
            <div data-role="footer" class="ui-bar">
                <a href="index.html" data-role="button" data- icon="delete">Remove</a>
                <a href="index.html" data-role="button" data-icon="plus">Add</a>
                <a href="index.html" data-role="button" data-icon="arrow-
u">Up</a>
                <a href="index.html" data-role="button" data-icon="arrow-
d">Down</a>
            </div>
        </div>
        <!-- First Page End -->
    </body>

</html>
```

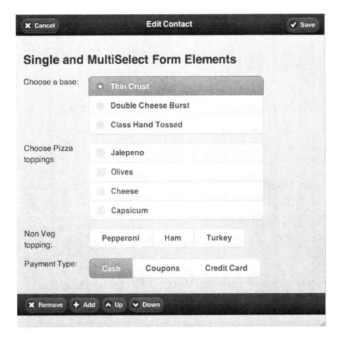

Figure 4–10. *jQueryMobile form elements*

List Views

So far, you have seen how easy it is to declare various UI widgets with simple HTML elements with data-roles and CSS classes. Well, in jQueryMobile, a list view is no exception. The following example will show you how a easily an HTML list can be converted to a mobile scrollable list.

```
<ul data-role="listview" data-theme="g">
    <li>
        <a href="usa.HTML">USA</a>
    </li>
    <li>
        <a href="uk.HTML">UK</a>
    </li>
    <li>
        <a href="russia.HTML">Russia</a>
    </li>
</ul>
```

The following is the complete code example of how to declare a list in HTML. In case of dynamic data, all you have to do is append the li elements within the ul element at runtime. Refer to Figure 4–11 to see what the list view looks like in jQueryMobile.

```
<!DOCTYPE HTML>
<HTML>

    <head>
        <title>jQuery Mobile Demo</title>
```

```html
        <link href="jquery.mobile-1.0rc2.min.css" rel="stylesheet" type="text/css"/>
        <script src="jquery-1.6.4.min.js"></script>
        <script src="jquery.mobile-1.0rc2.min.js"></script>
    </head>

    <body>
        <!-- Main Page-->
        <div data-role="page" id="main">
            <div data-role="header">
                <h1>
                    Header
                </h1>
            </div>
            <div data-role="content">
                <ul data-role="listview" data-theme="c">
                    <li>
                        <a href="usa.HTML">USA</a>
                    </li>
                    <li>
                        <a href="uk.HTML">UK</a>
                    </li>
                    <li>
                        <a href="russia.HTML">Russia</a>
                    </li>
                </ul>
            </div>
            <div data-role="footer" class="ui-bar">
                <h1>
                    Footer
                </h1>
            </div>
        </div>
        <!-- First Page End -->
    </body>

</HTML>
```

Figure 4-11. *jQueryMobile list view*

You can read more about this at the jQueryMobile website. See the demo and documentation by following this link: http://jquerymobile.com/demos/1.0a4.1/.

jQueryMobile Event Handling

In jQueryMobile, event handling can be categorized into two areas:

1. Events generated by non-jQueryMobile widgets. Examples of these are textboxes, text areas, buttons, radio buttons, and so on.

2. Events generated by jQueryMobile widgets and framework. Examples of these are the touch event, orientation change event, scroll events, and page lifecycle events.

Normal Events

Normal events should be handled as jQuery typically does. In jQuery, we have a generic method bind available with jQuery selector, which allows us to bind to any event.

```
<html>
    <head>…</head>
    <body>
        <button id="mybutton">mybutton</button>
    </body>
</html>

$("#mybutton").bind("click",function(event){
    alert("clicked mybutton");
});
```

jQuery also provides many convenience method for events like a short hand of above method is click method, which binds the callback to click event.

```
$("#mybutton").click(function(event){
    alert("clicked mybutton");
});
```

You can read more about jQuery events at the following website:

http://api.jquery.com/category/events/

Now, coming to the events generated by jQueryMobile. Please note that all events, irrespective of their source, need to be handled in the manner depicted above. The same applies to the events documented below.

The following section will discuss the various events generated by jQueryMobile framework and widgets.

Touch Events

Touch events on a mobile or tablet are very different than the traditional mouse events like click or double click. In a similar manner, in traditional mouse events, gestures are not possible. jQueryMobile provides a set of new events meant for touch gestures. These events are described in Table 4-2.

Table 4–2. *jQueryMobile Touch Events*

Event Name	Description
Tap	This event is generated after a quick touch and lift.
Taphold	This is a tap event where the finger (or any object) is held against the screen as in a long press for approximately one second.
Swipe	This is a touch gesture in which there is a horizontal drag of at least 30 px in any direction, limiting the vertical movement within 20 px. All this happens in one second.
Swipeleft	This is a swipe event but when the direction of swipe is toward the left.
Swiperight	This is a swipe event but when the direction of swipe is toward the right.

The following is an example of how touch events work with jQueryMobile. Refer to Figure 4–12 for this example.

```html
<!DOCTYPE html>
<html>

    <head>
        <title>jQuery Mobile Touch Events Demo</title>
        <link href="jquery.mobile-1.0rc2.min.css" rel="stylesheet" type="text/css"/>
        <script src="jquery-1.6.4.min.js"></script>
        <script src="jquery.mobile-1.0rc2.min.js"></script>
        <script>
            $(document).ready(function() {
                $("#tap").bind("tap", function() {
                    alert("TapEvent");
                });
                $("#taphold").bind("taphold", function() {
                    alert("Tap Hold Event");
                });
                $("#swipe").bind("swipe", function() {
                    alert("Swipe Event");
                });
                $("#swipeleft").bind("swipeleft", function() {
                    alert("Swipe Left Event");
                });
                $("#swiperight").bind("swiperight", function() {
                    alert("Swipe Right Event");
                });
            });
        </script>
    </head>

    <body>
        <!--Main Page-->
        <div data-role="page" id="main">
            <div data-role="header">
                <h1>
                    Touch Events
                </h1>
            </div>
            <div data-role="content">
                <h1>
                    Touch Events example
                </h1>
                <p id="tap">
                    Tap here
                </p>
                <p id="taphold">
                    Tap and hold here
                </p>
                <p id="swipe">
                    Swipe in this area.
                </p>
                <p id="swipeleft">
                    Swipe Left &lt;-- in this area.
                </p>
```

```
                <p id="swiperight">
                        Swipe Right -- &gt; in this area.
                </p>
            </div>
            <div data-role="footer" class="ui-bar">
                <h1>
                        Footer
                </h1>
            </div>
        </div>
        <!-- First Page End -->
    </body>

</html>
```

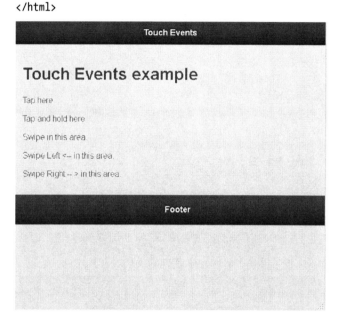

Figure 4–12. *jQueryMobile touch events.*

Orientation Change Events

Both mobile and tablet devices can detect and react to orientation change. This is very useful, as the aspect ratio of mobile or tablet is different in portrait mode and landscape mode. jQueryMobile allows the developer to use this orientation change to make the best use of real estate on the screen in both modes. In order to do this, you need to listen to the "orientationchange" event on the window element. This is shown in the following example. Refer to Figure 4–13 and 4–14.

```
<!DOCTYPE html>
<html>

    <head>
        <title>jQuery Mobile Touch Events Demo</title>
        <link href="jquery.mobile-1.0rc2.min.css" rel="stylesheet" type="text/css"/>
```

```
<script src="jquery-1.6.4.min.js"></script>
<script src="jquery.mobile-1.0rc2.min.js"></script>
    $(document).ready(function(){
        $(window).bind('orientationchange', function(event){
            $("#placeholder").html("Orientation changed to "+event.orientation);
        });
    });
</script>
</head>

<body>
    <!-- Main Page-->
    <div data-role="page" id="main">
        <div data-role="header">
            <h1>
                Touch Events
            </h1>
        </div>.
        <div data-role="content">
            <h1>
                Orientation Events example
            </h1>
            <div id="placeholder">
            </div>
        </div>
        <div data-role="footer" class="ui-bar">
            <h1>
                Footer
            </h1>
        </div>
    </div>
    <!-- First Page End -->
</body>

</html>
```

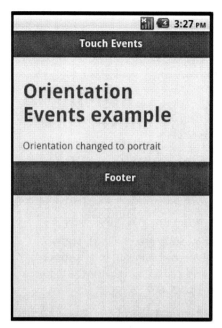

Figure 4–13. *jQueryMobile orientation change event*

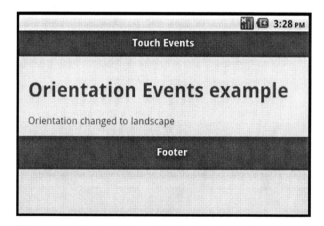

Figure 4–14. *jQueryMobile orientation change event*

Scroll Events

One of the important aspects of mobile devices is the ability to scroll and do things in the background when the scroll is happening. Think about lazy loaded lists, which fetch data as the user scrolls the list.

In order to do this, you need scroll events (see Table 4–3). jQueryMobile provides scroll events. Please note when the scrollstart event does not work as expected on iOS and we advise not to rely on scrollstart event on iOS.

Table 4–3. *jQueryMobile Scroll Events*

Event Name	Description
scrollstart	This event is fired when scroll starts. Note this event is not fired properly on iPhone. This is because iOS devices freeze DOM manipulation. All events during that time are queued and fired at the time scroll stops.
scrollstop	This event is fired when scroll stops.

Page Events

jQueryMobile has the concept of pages. Pages in jQueryMobile are created, shown, and/or hidden. jQueryMobile provides events so that developer can put in appropriate handling before the page is created, after it is created, and before and after the page is shown and hidden. All these events are documented in Table 4–4.

Table 4–4. *jQueryMobile Page Events*

Event Name	Description
pagebeforecreate	This event is fired just before the page is created.
pagecreate	This event is fired after the page is created.
pagebeforeshow	This event is fired before the page is shown.
pagebeforehide	This event is fired before the page is hidden.
pageshow	This event is fired after the page is shown.
pagehide	This event is fired after the page is hidden.

PhoneGap jQueryMobile Integration

Now that you understand how jQueryMobile works, let's work on integrating the features of PhoneGap and jQueryMobile to create applications.

Please note when you are using jQueryMobile with PhoneGap, there are three JavaScript frameworks, each bootstrapping on their own.

1. PhoneGap framework

2. jQuery framework

3. jQueryMobile framework

While all the frameworks provide their own bootstrap mechanism, it is best to bootstrap these frameworks in the following order:

1. PhoneGap

2. jQuery

3. jQueryMobile (if really required).

This is shown in the following example:

```
<script>

    //onDeviceReady is called when PhoneGap is initialized

    function onDeviceReady() {
        $(document).ready(function() {
            //Call any jQuery functions here
        });
    }
    document.addEventListener("deviceready", onDeviceReady);
</script>
```

Local Search Using jQueryMobile and PhoneGap

Let's put jQueryMobile and PhoneGap to the test. You will build a mashup of PhoneGap geo, compass, and a database feature combined with Google Maps Places API with the UI built using jQueryMobile. Refer to Figure 4–15.

This mashup is called a local search. This mashup has the following features:

1. Allows a user to search for a place of interest in the given radius of his/her current location.

2. Allows a user to look at the details of the place and also visit the website of the place.

3. Allows a user to save a place as a favorite and remove it from the favorites.

4. Allows a user to browse the favorites he has saved.

5. Allows a user to look at all the search result places in a Google Map.

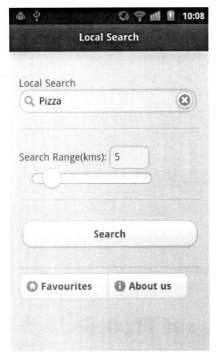

Figure 4–15. *Local search using jQueryMobile and PhoneGap*

The two most important features in this screen are as follows (refer to Figure 4–16):

1. The search button to search pizza in a 5 km radius of the user's location

2. The favorites button to show the user all the places marked as favorite

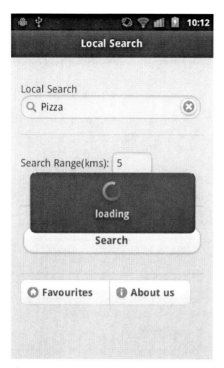

Figure 4–16. *Local search using jQueryMobile and PhoneGap*

The search results can be show as a list or as markers on a map. Figure 4–17shows how the results are seen as a list in.

Figure 4–17. *Local search result*

When the user clicks on one of the search results, he is taken to the business details page. This page shows details such as name, address and phone number, and vicinity of the business. It also allows the user to either add the business entry to his/her favorites or remove it from the favorites. This is shown in Figure 4–18.

Figure 4–18. *Local search business detail*

When a user adds a business to his favorites, he can navigate to the favorites page from the main page. Here, the user will see the places/business saved as favorites. These entries are actually stored to the app's internal database (provided by PhoneGap)(see Figure 4–19).

Figure 4–19. *Locally stored favorites .*

Here again, when the user clicks on one of the entries, he is taken to the details page. If you can observe, we can now see the "Remove to favorite" button, as this business/place is already part of the favorites. Refer to Figure 4–20.

Figure 4–20. *Favorite detail.*

Last, but not the least, is the entire search results plotted on a Google Map. In order to do so, you need to go to the main page, hit search, and then click on the map tab. You will see a screen as depicted in Figure 4–21.

Figure 4–21. *Local search result on a map*

Bootstrapping PhoneGap and jQuery

The bootstrapping is done in the following order:

1. PhoneGap is bootstrapped to call onDeviceReady() function when the PhoneGap JavaScript library is called

2. In the onDeviceReady, jQuery is bootstrapped to call the anonymous function when jQuery is loaded

```
<script>
function onDeviceReady(){
            $(document).ready(function(){
                    //Register event handlers
            });
}
document.addEventListener("deviceready",onDeviceReady);
</script>
```

Installing Necessary JavaScript Libraries

For this project, you will need following JavaScript libraries

1. jQuery: http://docs.jquery.com/Downloading_jQuery#Download_jQuery

2.   jQueryMobile: `http://jquerymobile.com/download/`

3.   jQuery ui map: `http://code.google.com/p/jquery-ui-map/downloads/list`

4.   PhoneGap: `www.phonegap.com/download/`

Assuming your application JavaScript is called app.js and application CSS is named app.css, your www folder should look as shown in Figure 4–22. Note the images folder is of jQueryMobile library.

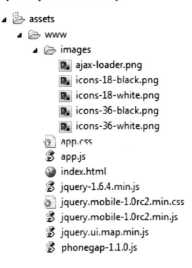

Figure 4–22. *Local search project structure*

Layout of Local Search

The local search page is the main page in the application. It takes three inputs to be able to do a local search.

1.   Geo location from PhoneGap

2.   Search keyword from the search text box

3.   Radius of search from the jQueryMobile slider

Once this is done, it will fetch the search results and show them in the search result page.

```
<!-- Main Search Page -->
<div data-role="page">
    <div data-role="header">
        <h1>
            Local Search
        </h1>
    </div>
    <!-- /header -->
    <div data-role="content">
```

```
        <div data-role="fieldcontain">
            <label for="search">
                Local Search
            </label>
            <input type="search" name="searchbox" id="searchbox" value="Pizza" />
        </div>
        <div data-role="fieldcontain">
            <label for="slider">
                Search Range(kms):
            </label>
            <input type="range" name="range" id="range" value="5" min="1" max="25"
            />
        </div>
        <div data-role="fieldcontain">
            <button name="search" id="search">
                Search
            </button>
        </div>
        <div data-role="controlgroup" data-type="horizontal">
            <a href="#fav" data-role="button" data-icon="home">Favorites</a>
            <a href="index.html" data-role="button" data-icon="info">About us</a>
        </div>
    </div>
    <!-- /content -->
</div>
```

Searching for a Local Business

In order to search for places of interest, we are using Google Maps Places API. Places of interest can be searched by making a restful call to the Google Maps Places service end point. There are many parameters required to make this restful call; these are shown in Table 4–5.

Table 4–5. *Google Place API Parameters*

Parameter name	Description
API key	You need an API key to tell Google that it is your application that is making rest calls. You need to register with Google and generate a key for your website. Visit this URL for more detail: http://code.google.com/apis/maps/signup.html
Latitude	Since Google Maps Places is a Geo Service, it requires your coordinates
Longitude	Since Google Maps Places is a Geo Service, it requires your coordinates
Radius	Google Maps Places needs a radius from you to determine the scope of the search
Name	The search keyword for finding places
Types	The category of places; this is "food" in this example

The URL is shown here with fillers for each parameter:

```
https://maps.googleapis.com/maps/api/place/search/json?location={latitude,longi
tude}&radius={radius}&types=food&name={search keyword}
&sensor=false&key={api_key}.
```

For the following URL, the JSON response occurs:

```
https://maps.googleapis.com/maps/api/place/search/json?location=-
33.8670522,151.1957362&radius=500&types=food&name=harbour&sensor=true&key=<<api
key>>.
```

JSON response is as follows:

```
{
    "html_attributions": ["Listings by \u003ca
href=\"http://www.yellowpages.com.au/\"\u003eYellow Pages\u003c/a\u003e"],
    "results": [{
        "geometry": {
            "location": {
                "lat": -33.8719640,
                "lng": 151.1985440
            }
        },
        "icon": "http://maps.gstatic.com/mapfiles/place_api/icons/restaurant-71.png",
        "id": "aefbc59325ffd5f3e93d67932375d20d143289de",
        "name": "Toros Restaurant Darling Harbour",
        "reference":
"CoQBdgAAAE6oRybc13OZYNHOWeuwKzTfzjYXO8nuWyGqCqSTBogR_BZxE3OfgXsybOl_wIROs_uuHLZqq-
17DTgpGHZoSehSbOG73dfIxO3rpQak2OmNuBb5Kg63rPN_afbH_PnbILiofw6WSODYOCkqhFl38qSXyujAPkQKZU
76NJypgT6mEhCg1MhyNAuyark4X8YfRg4YGhTn_MXrOgelHUHPe3JMCic-cHlu3A",
        "types": ["restaurant", "food", "establishment"],
        "vicinity": "Darling Dr, Sydney"
    }, ...]
}
```

Note that in the previous JSON, there is an id and reference. You need to understand these better to be able to build your application.

Id is a unique identifier of a place. You will use id as a primary key when you store the place/business in your database. However, id cannot be used to fetch the latest information.

Reference is a key that is used to fetch the details of a place/business from the Google Places server. However, note that a reference is not unique across multiple search results.

Therefore, you will store both id (as the primary key) and reference (as a string) in the database, so you can uniquely identify a place/business (using id) and fetch information from Google Places any time using the reference.

Overall Layout in HTML

The following is the overall layout of the application. There are in all five pages in the application:

1. Search page

2. Search result page with id "list"

3. Details page with id "DETAILS"

4. Favorite list page with id "fav"

5. Map page with id "map"

```
<!DOCTYPE HTML>
<html>

    <head>
        <title>PhoneGap</title>
        <link rel="stylesheet" type="text/css" href="app.css" />
        <script type="text/javascript"
src="http://maps.google.com/maps/api/js?sensor=true"></script>
        <script type="text/javascript" src="jquery.ui.map.min.js"></script>
        <link href="jquery.mobile-1.0rc2.min.css" rel="stylesheet" type="text/css"/>
        <script src="jquery-1.6.4.min.js"></script>
        <script src="jquery.mobile-1.0rc2.min.js"></script>
        <script type="text/javascript" src="phonegap-1.1.0.js"></script>
        <script type="text/javascript" src="app.js"></script>
    </head>

    <body>
        <!-- Main Search Page -->
        <div data-role="page">
            <div data-role="header">
                <h1>
                    Local Search
                </h1>
            </div>
            <!-- /header -->
            <div data-role="content">
                <div data-role="fieldcontain">
                    <label for="search">
                        Local Search
                    </label>
                    <input type="search" name="searchbox" id="searchbox" value="Pizza"
/>
                </div>
                <div data-role="fieldcontain">
                    <label for="slider">
                        Search Range(kms):
                    </label>
                    <input type="range" name="range" id="range" value="5" min="1"
max="25"
                    />
                </div>
```

```html
                        <div data-role="fieldcontain">
                            <button name="search" id="search">
                                Search
                            </button>
                        </div>
                        <div data-role="controlgroup" data-type="horizontal">
                            <a href="#fav" data-role="button" data-icon="home">Favorites</a>
                            <a href="index.html" data-role="button" data-icon="info">About
us</a>
                        </div>
                    </div>
                    <!-- /content -->
                </div>
                <!-- /page -->
                <!-- Search Result List Page -->
                <div data-role="page" id="list">
                    <div data-role="header" data-position="fixed">
                        <h1>
                            Result
                        </h1>
                    </div>
                    <!-- /header -->
                    <div data-role="content">
                        <ul id="result-list" data-role="listview" data-theme="g">
                        </ul>
                    </div>
                    <!-- /content -->
                    <div data-role="footer" data-id="result-footer" data-position="fixed"
                        class="ui-bar-a
                            ui-footer ui-footer-fixed fade ui-fixed-overlay"
                        role="contentinfo" style="top: -1263px; ">
                            <div data-role="navbar" class="ui-navbar ui-navbar-noicons"
role="navigation">
                                <ul class="ui-grid-a">
                                    <li class="ui-block-a">
                                        <a href="#list" data-theme="a"
                                            class="ui-btn-active ui-state-persist ui-btn ui-btn-up-
a">
                                            <span class="ui-btn-inner"><span class="ui-btn-
text">List</span></span>
                                        </a>
                                    </li>
                                    <li class="ui-block-b">
                                        <a href="#map" data-theme="a"
                                            class="ui-state-persist ui-btn ui-btn-up-a">
                                            <span class="ui-btn-inner"><span class="ui-btn-
text">Maps</span></span>
                                        </a>
                                    </li>
                                </ul>
                            </div>
                            <!-- /navbar -->
                    </div>
                    <!-- /footer -->
                </div>
                <!-- /page -->
                <!-- Maps Page -->
```

```
<div data-role="page" id="map">
    <div data-role="header">
        <h1>
            Map
        </h1>
    </div>
    <!-- /header -->
    <div data-role="content" class="map-content">
        <div id="map_canvas">
        </div>
    </div>
    <!-- /content -->
    <div data-role="footer" data-id="result-footer"
        data-position="fixed"
        class="ui-bar-a ui-footer ui-footer-fixed fade ui-fixed-overlay"
        role="contentinfo" style="top: -1263px; ">
        <div data-role="navbar" class="ui-navbar ui-navbar-noicons"
role="navigation">
            <ul class="ui-grid-a">
                <li class="ui-block-a">
                    <a href="#list" data-theme="a" class="ui-state-persist ui-
btn ui-btn-up-a">
                        <span class="ui-btn-inner"><span class="ui-btn-
text">List</span></span></a>
                </li>
                <li class="ui-block-b">
                    <a href="#map" data-theme="a"
                        class="ui-btn-active ui-state-persist ui-btn ui-btn-up-
a">
                        <span class="ui-btn-inner"><span class="ui-btn-
text">Maps</span></span></a>
                </li>
            </ul>
        </div>
        <!-- /navbar -->
    </div>
    <!-- /footer -->
</div>
<!-- /page -->
<!--Favorite List Page -->
<div data-role="page" id="fav">
    <div data-role="header">
        <h1>
            Favorites
        </h1>
    </div>
    <!-- /header -->
    <div data-role="content">
        <!--
            <ul id="fav-list" data-role="listview" data-theme="g"></ul>
        -->
        <ul id="fav-list" data-role="listview" data-theme="g">
        </ul>
    </div>
    <!-- /content -->
    <div data-role="footer" data-id="result-footer" data-position="fixed"
        class="ui-bar-a ui-footer ui-footer-fixed fade ui-fixed-overlay"
```

```
                              role="contentinfo" style="top: -1263px; ">
                              <!-- /navbar -->
                      </div>
                      <!-- /footer -->
              </div>
              <!-- /page -->
              <!-- Business Details Page -->
              <div data-role="page" id="details">
                  <div data-role="header">
                      <h1>
                              Business Details
                      </h1>
                  </div>
                  <!-- /header -->
                  <div data-role="content">
                      <table summary="Business Details">
                          <caption>
                              <h3>
                                      Business Details
                              </h3>
                          </caption>
                          <tfoot>
                              <tr>
                                  <td colspan="2">
                                      <div id="remove">
                                          <button id="removefav" data-role="button">
                                              Remove to Favorite
                                          </button>
                                      </div>
                                      <div id="add">
                                          <button id="addfav" data-role="button">
                                              Add to Favorite
                                          </button>
                                      </div>
                                  </td>
                              </tr>
                              <tr>
                                  <td colspan="2">
                                      <a id="homepage" data-role="button" href="">Visit
        HomePage</a>
                                  </td>
                              </tr>
                          </tfoot>
                          <tbody>
                              <tr>
                                  <th scope="row">
                                      Name
                                  </th>
                                  <td id="name">
                                      ...
                                  </td>
                              </tr>
                              <tr>
                                  <th scope="row">
                                      Address
                                  </th>
                                  <td id="address">
```

```
                             ...
                        </td>
                    </tr>
                    <tr>
                        <th scope="row">
                            Phone
                        </th>
                        <td id="phone">
                            ...
                        </td>
                    </tr>
                    <tr>
                        <th scope="row">
                            Rating
                        </th>
                        <td id="rating">
                            ...
                        </td>
                    </tr>
                </tbody>
            </table>
        </div>
        <!-- /content -->
    </div>
    <!-- /page -->
</body>

</html>
```

Fetching and Showing the Search Results

The function-initiated search binds the search button event with the function that actually performs the search.

The following is the flow of events for a search:

1. Show user a loading icon to let him/her know a long operation is occurring. This is done by calling $.mobile.showPageLoadingMsg();.

2. Get the current position of the user using PhoneGap function navigator.geolocation.getCurrentPosition(successCallback, failureCallback).

3. In the successCallback of the above call, a JSON request to Google Places with the following parameters is done:

   a. Geolocation

   b. Keyword for search

   c. Radius of search from the geolocation

   d. Developer Api key for Google Places

   e. `var url="https://maps.googleapis.com/maps/api/place/search/json?locati on="+position.coords.latitude+","+position.coords.longitude+"&radi`

```
us="+radius+"&name="+$("#searchbox").val()+"&sensor=true&key=<API_
Key>";
```

4. jQuery's $.getJSON() is used to make a Ajax call to Google Places to fetch a JSON response. Note that as the PhoneGap application has no domain, you are not restricted by single origin policy of browser. Register a successCallback and a failureCallback with $.getJSON() function.

5. In the successCallback of above call, the places response are fetched and appended to the ul element whose id is "result-list." This ul element is annotated in the HTML code as a jQueryMobile list view. Once you have added the necessary li element to the ul element, we will call $("result-list").listView("refresh") to redraw the ul element as a jQueryMobile list.

6. Note that you put the reference part of the JSON response for each place as the id of the link (an element). You do this so that when the user taps on this entry, he is taken to the details page. Note that the href part of this link is actually "#details."

7. Lastly, you are binding a click handler with the click event on the place so that you can actually make a call to the Google Places server to fetch the details of the business/place entry before navigating the user to the details page.

8. The last step is to remove the loading icon by calling $.mobile.hidePageLoadingMsg ();.

```
/**
 * Binding Search button handler to go and fetch place results
 */
function initiateSearch(){
        $("#search").click(function(){
try {
                $.mobile.showPageLoadingMsg();

navigator.geolocation.getCurrentPosition(function(position){

var radius = $("#range").val() * 1000;
mapdata = new google.maps.LatLng(position.coords.latitude, position.coords.longitude);
var url = "https://maps.googleapis.com/maps/api/place/search/json?location=" +
position.coords.latitude + "," + position.coords.longitude + "&radius=" + radius +
"&name=" + $("#searchbox").val() +
"&sensor=false&key=AIzaSyC4vCfT_Knq1SGuNMahZqyrmZFiTuBsdlY";
                $.getJSON(url, function(data){
cachedData = data;
                $("#result-list").html("");
try {
                        $(data.results).each(function(index, entry){

var htmlData = "<a href=\"#details\" id=\"" + entry.reference + "\"><img src=\"" +
entry.icon + "\" class=\"ui-li-icon\"></img><h3> " + entry.name +
```

```
                           "</h3><p><strong> vicinity:" + entry.vicinity + "</strong></p></a>";
                           var liElem = $(document.createElement('li'));

                                        $("#result-list").append(liElem.html(htmlData));


                                        $(liElem).bind("tap", function(event){
                           event.stopPropagation();
                           fetchDetails(entry);
                           return true;
                                        });

                                    });
                                                                $("#result-
                           list").listview('refresh');
                                    }
                           catch (err) {
                           console.log("Got error while putting search result on result page " + err);
                                    }


                                    $.mobile.changePage("list");
                                    $.mobile.hidePageLoadingMsg();
                               }).error(function(xhr, textStatus, errorThrown){
                           console.log("Got error while fetching search result : xhr.status=" + xhr.status);

                               }).complete(function(error){
                                    $.mobile.hidePageLoadingMsg();
                               });
                           }, function(error){
                           console.log("Got Error fetching geolocation " + error);
                                    });


                                }
                           catch (err) {
                           console.log("Got error on clicking search button " + err);
                                }



                        });

                    }
```

Showing Details of a Place/Business

You have seen above that when the user clicks on a place entry in the list the showing places search results, fetchDetails() function is called.

The fetchDetails() function has the following flow:

1. The user is shown a loading icon. This is done by calling $.mobile.showPageLoadingMsg().

2. All the fields of the details place are reset to be blank. For example, $("#name").html();.

3. A URL of the details place request is created (detailsURL) and an Ajax call is made using jQuery $.getJSON() call.

4. In the successCallback of $.getJSON(), you get the details of the page. Here you first check if the given place is already stored by a user as a favorite or not. Based on this, the user is shown an "add to favorite" or "remove from favorite" button.

5. The fields of this page are populated.

6. The loading icon is removed by calling $.mobile.hidePageLoadingMsg().

```
/**
    * Fetch the details of a place/business. This function is called before user
navigates to details page
    * @param {Object} reference
    */
function fetchDetails(entry){

currentBusinessData = null;

        $.mobile.showPageLoadingMsg();
var detailsUrl = "https://maps.googleapis.com/maps/api/place/details/json?reference=" +
entry.reference + "&sensor=true&key=<API_Key>";
        $("#name").html("");
        $("#address").html("");
        $("#phone").html("");
        $("#rating").html("");
        $("#homepage").attr("href", "");

        $.getJSON(detailsUrl, function(data){
if (data.result) {
currentBusinessData = data.result;

isFav(currentBusinessData, function(isPlaceFav){
                                console.log(currentBusinessData.name+" is fav
"+isPlaceFav);
if (!isPlaceFav) {

                    $("#add").show();
                    $("#remove").hide();
                }
else {

                    $("#add").hide();
                    $("#remove").show();
                }
                $("#name").html(data.result.name);
                $("#address").html(data.result.formatted_address);
                $("#phone").html(data.result.formatted_phone_number);
                $("#rating").html(data.result.rating);
                $("#homepage").attr("href", data.result.url);

            });
```

```
        }
    }).error(function(err){
console.log("Got Error while fetching details of Business " + err);
    }).complete(function(){
        $.mobile.hidePageLoadingMsg();
    });

}
```

Adding and Removing a Place/Business to Favorite

The next step is to actually see

1. How do we add a place to our favorites list?

2. How do we remove a place from our favorites list?

3. How do we find a given place is part of our favorites list?

The thing to note here is that you are using PhoneGap's database API to store, retrieve, and delete places. All this information is stored in the application's database using PhoneGap's database API. You will store the favorites in a table named "favorite."

The initiateFavButton() binds the click handlers for the "add to favorite" and "remove from favorite" buttons to actual handler. The "add to favorite" button is within a div with id "add," and "remove from favorite" is in a div with id "remove." You are controlling the visibility of the buttons by hiding or showing these divs. You also give corresponding calls to addToFavorite() and removeFromFavorite() methods to actually do the add and remove.

```
/**
 * Called to bind the "Add to Favorite" Button
 */

function initiateFavButton() {
    $("#removefav").click(function () {

        try {
            if (currentBusinessData != null) {
                removeFromFavorite(currentBusinessData);
                $("#add").show();
                $("#remove").hide();

            }
        } catch (err) {
            console.log("Got Error while removing " + currentBusinessData.name + " error
" + err);
        }


    });
    $("#addfav").click(function () {
```

```
            try {
                if (currentBusinessData != null) {
                    addToFavorite(currentBusinessData);
                    $("#add").hide();
                    $("#remove").show();
                }
            } catch (err) {
                console.log("Got Error while adding " + currentBusinessData.name + " error "
+ err);
            }

        });


}
```

The ensureTableExists() is a common function used by all other database functions. This function ensures you execute the SQL script "CREATE TABLE IF NOT EXISTS Favorite (id unique, reference, name, address, phone, rating, icon, vicinity)" before doing any insert, select, or delete operations on the database.

```
/**
 * Ensure we have the table before we use it
 * @param {Object} tx
 */

function ensureTableExists(tx) {
    tx.executeSql('CREATE TABLE IF NOT EXISTS Favorite (id unique, reference,
name,address,phone,rating,icon,vicinity)');
}
```

The addToFavorite() is the function that actually does the database insert for the given places in the favorites table. Note that you are storing id, reference, name, icon, formatted_address, formatted_phone_number, rating, and vicinity in the database table "favorite".

```
    /**
     * Add current business data to favorite
     * @param {Object} data
     */

    function addToFavorite(data) {
        var db = window.openDatabase("Favorites", "1.0", "Favorites", 20000000);

        db.transaction(function (tx) {
            ensureTableExists(tx);
            var id = (data.id != null) ? ('"' + data.id + '"') : ('""');
            var reference = (data.reference != null) ? ('"' + data.reference + '"') :
('""');
            var name = (data.name != null) ? ('"' + data.name + '"') : ('""');
            var address = (data.formatted_address != null) ? ('"' +
data.formatted_address + '"') : ('""');
            var phone = (data.formatted_phone_number != null) ? ('"' +
data.formatted_phone_number + '"') : ('""');
            var rating = (data.rating != null) ? ('"' + data.rating + '"') : ('""');
            var icon = (data.icon != null) ? ('"' + data.icon + '"') : ('""');
```

```
          var vicinity = (data.vicinity != null) ? ('"' + data.vicinity + '"') :
('""');
          var insertStmt = 'INSERT INTO Favorite (id,reference,
name,address,phone,rating,icon,vicinity) VALUES (' + id + ',' + reference + ',' + name +
',' + address + ',' + phone + ',' + rating + ',' + icon + ',' + vicinity + ')';
          tx.executeSql(insertStmt);

      }, function (error) {
          console.log("Data insert failed " + error.code + "    " + error.message);
      }, function () {
          console.log("Data insert successful");
      });

  }
```

The removeFromFavorite() is the function that removes the favorite from the "favorite" table. It only requires the id to do so.

```
/**
 * Remove current business data from favorite
 * @param {Object} data
 */

function removeFromFavorite(data) {
    try {
        var db = window.openDatabase("Favorites", "1.0", "Favorites", 20000000);

        db.transaction(function (tx) {
            ensureTableExists(tx);
            var deleteStmt = "DELETE FROM Favorite WHERE id = '" + data.id + "'";
            console.log(deleteStmt);
            tx.executeSql(deleteStmt);

        }, function (error) {
            console.log("Data Delete failed " + error.code + "    " + error.message);
        }, function () {
            console.log("Data Delete successful");
        });
    } catch (err) {
        console.log("Caught exception while deleting favorite " + data.name);
    }

}
```

The isFav() is the function that queries the table "favorite" to find out whether the given place/business is already present in the table and thus marked by the user as a favorite.

```
    /**
     *
     * @param {Object} reference
     * @return true if place is favorite else false
     */

    function isFav(data, callback) {
        var db = window.openDatabase("Favorites", "1.0", "Favorites", 200000);
        try {
            db.transaction(function (tx) {
                ensureTableExists(tx);
```

```
                        var sql = "SELECT * FROM Favorite where id='" + data.id + "'";
                        tx.executeSql(sql, [], function (tx, results) {

                            var result = (results != null && results.rows != null &&
      results.rows.length > 0);

                            callback(result);
                        }, function (tx, error) {

                            console.log("Got error in isFav error.code =" + error.code + "
      error.message = " + error.message);
                            callback(false);


                        });
                    });
                } catch (err) {
                    console.log("Got error in isFav " + err);
                    callback(false);
                }
      }
```

Loading Your Favorite Places

So far you have added and removed a place to and from the favorites. You also saw
how to check whether a place is set by the user as his/her favorite. Now, let's look at the
code that retrieves all the favorite places of a user.

This code is called when the user hits the "favorites" button on the home page.

This code is very similar to isFav(), except here you will actually fetch all the entries from
the "favorite" table, take the result set, and populate the UL with "fav-list" id.

The populating part is similar to showing the results of a search.

Please make a note that you fetch results from the database table "favorites" each time
the user navigates to "favorites place." This is done by listening to "pagebeforeshow"
event of the page. The "pagebeforeshow" event is fired just before the jQueryMobile
page is shown to the user.

```
/**
 * Called each time before user navigates to Favorites
 */

function initiateFavorites() {
    $("#fav").live("pagebeforeshow", function () {

        var db = window.openDatabase("Favorites", "1.0", "Favorites", 200000);
        try {
            db.transaction(function (tx) {
                tx.executeSql('SELECT * FROM Favorite', [], function (tx, results) {

                    $("#fav-list").html("");
                    if (results != null && results.rows != null) {
                        for (var index = 0; index < results.rows.length; index++) {
```

```
                                var entry = results.rows.item(index)

                                var htmlData = "<a href=\"#details\" id=\"" +
entry.reference + "\"><img src=\"" + entry.icon + "\" class=\"ui-li-
icon\"></img><h3> " + entry.name + "</h3><p><strong> vicinity:" +
entry.vicinity + "</strong></p></a>";

                                var liElem = $(document.createElement('li'));

                                $("#fav-list").append(liElem.html(htmlData));

                                $(liElem).bind("tap", function (event) {
                                    event.stopPropagation();
                                    fetchDetails(entry);
                                    return true;
                                });

                            }
                            $("#fav-list").listview('refresh');
                        }
                    }, function (error) {
                        console.log("Got error fetching favorites " + error.code + " " +
error.message);
                    });
                });
            } catch (err) {
                console.log("Got error while reading favorites " + err);
            }

        });
}
```

Showing Search Result on a Map

The last part of this exercise is to show the search result for a place on Google Maps.
This helps the user get a better idea of where a place is located. In the initiateSearch()
function, when you received the search result, you cached the result in a JavaScript
variable named "cachedData". In this, you will actually use the same data to plot
markers on the map.

Please make a note that you redraw the map and plot the markers from cachedData
each time the user navigates to "favorites place." This is done by listening to the
"pagebeforeshow" event of the page.

```
/**
 * Called to initiate Map page
 */

function initiateMap() {
    $("#map").live("pagebeforecreate", function () {
        try {

            $('#map_canvas').gmap({
                'center': mapdata,
```

```
                            'zoom': 12,
                            'callback': function (map) {
                                $(cachedData.results).each(function (index, entry) {
                                    $('#map_canvas').gmap('addMarker', {
                                        'position': new
                                        google.maps.LatLng(entry.geometry.location.lat,
entry.geometry.location.lng),
                                        'animation': google.maps.Animation.DROP
                                    }, function (map, marker) {
                                        $('#map_canvas').gmap('addInfoWindow', {
                                            'position': marker.getPosition(),
                                            'content': entry.name
                                        }, function (iw) {
                                            $(marker).click(function () {
                                                iw.open(map, marker);
                                                map.panTo(marker.getPosition());
                                            });
                                        });
                                    });
                                });

                            });
                        }



                });
                console.log("Map initialized");
            } catch (err) {
                console.log("Got error while initializing map " + err);
            }

        });
```

Complete Source Code

The complete source of the index.html is as follows:

```
<!DOCTYPE HTML>
<html>

    <head>
        <title>PhoneGap</title>
        <link rel="stylesheet" type="text/css" href="app.css" />
        <script type="text/javascript"
src="http://maps.google.com/maps/api/js?sensor=true"></script>
        <script type="text/javascript" src="jquery.ui.map.min.js"></script>
        <link href="jquery.mobile-1.0rc2.min.css" rel="stylesheet" type="text/css"/>
        <script src="jquery-1.6.4.min.js"></script>
        <script src="jquery.mobile-1.0rc2.min.js"></script>
        <script type="text/javascript" src="phonegap-1.1.0.js"></script>
        <script type="text/javascript" src="app.js"></script>
    </head>

    <body>
        <!-- Main Search Page -->
```

```
        <div data-role="page">
            <div data-role="header">
                <h1>
                    Local Search
                </h1>
            </div>
            <!-- /header -->
            <div data-role="content">
                <div data-role="fieldcontain">
                    <label for="search">
                        Local Search
                    </label>
                    <input type="search" name="searchbox" id="searchbox" value="Pizza"
/>
                </div>
                <div data-role="fieldcontain">
                    <label for="slider">
                        Search Range(kms):
                    </label>
                    <input type="range" name="range" id="range" value="5" min="1"
max="25"
                    />
                </div>
                <div data-role="fieldcontain">
                    <button name="search" id="search">
                        Search
                    </button>
                </div>
                <div data-role="controlgroup" data-type="horizontal">
                    <a href="#fav" data-role="button" data-icon="home">Favorites</a>
                    <a href="index.html" data-role="button" data-icon="info">About
us</a>
                </div>
            </div>
            <!-- /content -->
        </div>
        <!-- /page -->
        <!-- Search Result List Page -->
        <div data-role="page" id="list">
            <div data-role="header" data-position="fixed">
                <h1>
                    Result
                </h1>
            </div>
            <!-- /header -->
            <div data-role="content">
                <ul id="result-list" data-role="listview" data-theme="g">
                </ul>
            </div>
            <!-- /content -->
            <div data-role="footer" data-id="result-footer" data-position="fixed"
            class="ui-bar-a
ui-footer ui-footer-fixed fade ui-fixed-overlay" role="contentinfo" style="top: -
1263px; ">
                <div data-role="navbar" class="ui-navbar ui-navbar-noicons"
role="navigation">
                    <ul class="ui-grid-a">
```

```
                    <li class="ui-block-a">
                        <a href="#list" data-theme="a" class="ui-btn-active ui-
state-persist ui-btn ui-btn-up-
a"><span class="ui-btn-inner"><span class="ui-btn-text">List</span></span></a>
                    </li>
                    <li class="ui-block-b">
                        <a href="#map" data-theme="a" class="ui-state-persist ui-btn
ui-btn-up-a"><span
class="ui-btn-inner"><span class="ui-btn-text">Maps</span></span></a>
                    </li>
                </ul>
            </div>
            <!-- /navbar -->
        </div>
        <!-- /footer -->
    </div>
    <!-- /page -->
    <!-- Maps Page -->
    <div data-role="page" id="map">
        <div data-role="header">
            <h1>
                Map
            </h1>
        </div>
        <!-- /header -->
        <div data-role="content" class="map-content">
            <div id="map_canvas">
            </div>
        </div>
        <!-- /content -->
        <div data-role="footer" data-id="result-footer" data-position="fixed"
        class="ui-bar-a
ui-footer ui-footer-fixed fade ui-fixed-overlay" role="contentinfo" style="top: -
1263px; ">
            <div data-role="navbar" class="ui-navbar ui-navbar-noicons"
role="navigation">
                <ul class="ui-grid-a">
                    <li class="ui-block-a">
                        <a href="#list" data-theme="a" class="ui-state-persist ui-
btn ui-btn-up-a"><span
class="ui-btn-inner"><span class="ui-btn-text">List</span></span></a>
                    </li>
                    <li class="ui-block-b">
                        <a href="#map" data-theme="a" class="ui-btn-active ui-state-
persist ui-btn ui-btn-up-
a"><span class="ui-btn-inner"><span class="ui-btn-text">Maps</span></span></a>
                    </li>
                </ul>
            </div>
            <!-- /navbar -->
        </div>
        <!-- /footer -->
    </div>
    <!-- /page -->
    <!-- Favorite List Page -->
    <div data-role="page" id="fav">
        <div data-role="header">
```

```
                <h1>
                    Favorites
                </h1>
            </div>
            <!-- /header -->
            <div data-role="content">
                <!-- <ul id="fav-list" data-role="listview" data-theme="g">
</ul>
                -->
                <ul id="fav-list" data-role="listview" data-theme="g">
                </ul>
            </div>
            <!-- /content -->
            <div data-role="footer" data-id="result-footer" data-position="fixed"
            class="ui-bar-a
ui-footer ui-footer-fixed fade ui-fixed-overlay" role="contentinfo" style="top: -
1263px; ">
                <!-- /navbar -->
            </div>
            <!-- /footer -->
        </div>
        <!-- /page -->
        <!-- Business Details Page -->
        <div data-role="page" id="details">
            <div data-role="header">
                <h1>
                    Business Details
                </h1>
            </div>
            <!-- /header -->
            <div data-role="content">
                <table summary="Business Details">
                    <caption>
                        <h3>
                            Business Details
                        </h3>
                    </caption>
                    <tfoot>
                        <tr>
                            <td colspan="2">
                                <div id="remove">
                                    <button id="removefav" data-role="button">
                                        Remove to Favorite
                                    </button>
                                </div>
                                <div id="add">
                                    <button id="addfav" data-role="button">
                                        Add to Favorite
                                    </button>
                                </div>
                            </td>
                        </tr>
                        <tr>
                            <td colspan="2">
                                <a id="homepage" data-role="button" href="">Visit
HomePage</a>
                            </td>
```

```
                            </tr>
                        </tfoot>
                        <tbody>
                            <tr>
                                <th scope="row">
                                    Name
                                </th>
                                <td id="name">
                                    ...
                                </td>
                            </tr>
                            <tr>
                                <th scope="row">
                                    Address
                                </th>
                                <td id="address">
                                    ...
                                </td>
                            </tr>
                            <tr>
                                <th scope="row">
                                    Phone
                                </th>
                                <td id="phone">
                                    ...
                                </td>
                            </tr>
                            <tr>
                                <th scope="row">
                                    Rating
                                </th>
                                <td id="rating">
                                    ...
                                </td>
                            </tr>
                        </tbody>
                    </table>
                </div>
                <!-- /content -->
            </div>
            <!-- /page -->
        </body>

</html>
```

The complete source of the app.js is as follows. Note that you will need to replace the
<API_Key> with your own key. You can get API Key from
http://code.google.com/apis/maps/documentation/places.

```
var mapdata = null;

var cachedData = null;

var currentBusinessData = null;

/**
    * Fetch the details of a place/business. This function is called before user
```

```
navigates to details page
    * @param {Object} reference
    */

function fetchDetails(entry) {

    currentBusinessData = null;

    $.mobile.showPageLoadingMsg();
    var detailsUrl =
"https://maps.googleapis.com/maps/api/place/details/json?reference=" + entry.reference +
"&sensor=true&key=<API_Key>";
    $("#name").html("");
    $("#address").html("");
    $("#phone").html("");
    $("#rating").html("");
    $("#homepage").attr("href", "");

    $.getJSON(detailsUrl, function (data) {
        if (data.result) {
            currentBusinessData = data.result;

            isFav(currentBusinessData, function (isPlaceFav) {
                console.log(currentBusinessData.name + " is fav
" + isPlaceFav);
                if (!isPlaceFav) {

                    $("#add").show();
                    $("#remove").hide();
                } else {

                    $("#add").hide();
                    $("#remove").show();
                }
                $("#name").html(data.result.name);
                $("#address").html(data.result.formatted_address);
                $("#phone").html(data.result.formatted_phone_number);
                $("#rating").html(data.result.rating);
                $("#homepage").attr("href", data.result.url);

            });



        }
    }).error(function (err) {
        console.log("Got Error while fetching details of Business " + err);
    }).complete(function () {
        $.mobile.hidePageLoadingMsg();
    });

}

//-------------------------------
/**
 * Called to initiate Map page
 */
```

```
function initiateMap() {
    $("#map").live("pagebeforecreate", function () {
        try {

            $('#map_canvas').gmap({
                'center': mapdata,
                'zoom': 12,
                'callback': function (map) {
                    $(cachedData.results).each(function (index, entry) {
                        $('#map_canvas').gmap('addMarker', {
                            'position': new
google.maps.LatLng(entry.geometry.location.lat, entry.geometry.location.lng),
                            'animation': google.maps.Animation.DROP
                        }, function (map, marker) {
                            $('#map_canvas').gmap('addInfoWindow', {
                                'position': marker.getPosition(),
                                'content': entry.name
                            }, function (iw) {
                                $(marker).click(function () {

                                    iw.open(map, marker);

                                    map.panTo(marker.getPosition());
                                });
                            });
                        });
                    });

                }

            });
            console.log("Map initialized");
        } catch (err) {
            console.log("Got error while initializing map " + err);
        }

    });


}
//-----------------------------------------------------------------------------
/**
 * Called to bind the "Add to Favorite" Button
 */

function initiateFavButton() {
    $("#removefav").click(function () {

        try {
            if (currentBusinessData != null) {
                removeFromFavorite(currentBusinessData);
                $("#add").show();
                $("#remove").hide();
```

```
                }
            } catch (err) {
                console.log("Got Error while removing " + currentBusinessData.name + " error
" + err);
            }



    });
    $("#addfav").click(function () {
        try {
            if (currentBusinessData != null) {

                addToFavorite(currentBusinessData);
                $("#add").hide();
                $("#remove").show();
            }
        } catch (err) {
            console.log("Got Error while adding " + currentBusinessData.name + " error "
+ err);
        }

    });



}
//-------------------------------------------------------------------------------------
-----------------
/**
 * Called each time before user navigates to Favorites
 */

function initiateFavorites() {
    $("#fav").live("pagebeforeshow", function () {

        var db = window.openDatabase("Favorites", "1.0", "Favorites", 200000);
        try {
            db.transaction(function (tx) {
                tx.executeSql('SELECT * FROM Favorite', [], function (tx, results) {

                    $("#fav-list").html("");
                    if (results != null && results.rows != null) {
                        for (var index = 0; index < results.rows.length;
                        index++) {

                            var entry = results.rows.item(index)

                            var htmlData = "<a href=\"#details\" id=\"" +
entry.reference + "\"><img src=\"" + entry.icon + "\" class=\"ui-li-
icon\"></img><h3> " + entry.name + "</h3><p><strong> vicinity:" +
entry.vicinity + "</strong></p></a>";

                            var liElem = $(document.createElement('li'));

                            $("#fav-
list").append(liElem.html(htmlData));
```

```
                                    $(liElem).bind("tap", function (event) {
                                        event.stopPropagation();
                                        fetchDetails(entry);
                                        return true;
                                    });

                                }
                                $("#fav-
list").listview('refresh');
                            }
                        }, function (error) {
                            console.log("Got error fetching Favorites " + error.code + " " +
error.message);
                        });
                    });
            } catch (err) {
                console.log("Got error while reading Favorites " + err);
            }

        });
    }
//-----------------------------------------------------------------------------
/**
 * Ensure we have the table before we use it
 * @param {Object} tx
 */

function ensureTableExists(tx) {
    tx.executeSql('CREATE TABLE IF NOT EXISTS Favorite (id unique,
reference, name,address,phone,rating,icon,vicinity)');
}
//-----------------------------------------------------------------------
/**
 * Add current business data to Favorite
 * @param {Object} data
 */

function addToFavorite(data) {
    var db = window.openDatabase("Favorites", "1.0", "Favorites", 20000000);

    db.transaction(function (tx) {
        ensureTableExists(tx);
        var id = (data.id != null) ? ('"' + data.id + '"') : ('""');
        var reference = (data.reference != null) ? ('"' + data.reference + '"') :
('""');
        var name = (data.name != null) ? ('"' + data.name + '"') : ('""');
        var address = (data.formatted_address != null) ? ('"' + data.formatted_address +
'"') : ('""');
        var phone = (data.formatted_phone_number != null) ? ('"' +
data.formatted_phone_number + '"') : ('""');
        var rating = (data.rating != null) ? ('"' + data.rating + '"') : ('""');
        var icon = (data.icon != null) ? ('"' + data.icon + '"') : ('""');
        var vicinity = (data.vicinity != null) ? ('"' + data.vicinity + '"') : ('""');
        var insertStmt = 'INSERT INTO Favorite (id,reference,
name,address,phone,rating,icon,vicinity) VALUES (' + id + ',' + reference + ',' + name +
',' + address + ',' + phone + ',' + rating + ',' + icon + ',' + vicinity + ')';
        tx.executeSql(insertStmt);
```

```
    }, function (error) {
        console.log("Data insert failed " + error.code + "    " + error.message);
    }, function () {
        console.log("Data insert successful");
    });

}
//-----------------------------------------------------------------------------
/**
 * Remove current business data from Favorite
 * @param {Object} data
 */

function removeFromFavorite(data) {
    try {
        var db = window.openDatabase("Favorites", "1.0", "Favorites", 20000000);

        db.transaction(function (tx) {
            ensureTableExists(tx);
            var deleteStmt = "DELETE FROM Favorite WHERE id
= '" + data.id + "'";
            console.log(deleteStmt);
            tx.executeSql(deleteStmt);

        }, function (error) {
            console.log("Data Delete failed " + error.code + "    " + error.message);
        }, function () {
            console.log("Data Delete successful");
        });
    } catch (err) {
        console.log("Caught exception while deleting Favorite " + data.name);
    }

}

//-------------------------------------------------------------------
/**
 *
 * @param {Object} reference
 * @return true if place is Favorite else false
 */

function isFav(data, callback) {
    var db = window.openDatabase("Favorites", "1.0", "Favorites", 200000);
    try {
        db.transaction(function (tx) {
            ensureTableExists(tx);
            var sql = "SELECT * FROM Favorite where id='" + data.id + "'";
            tx.executeSql(sql, [], function (tx, results) {

                var result = (results != null && results.rows != null &&
results.rows.length > 0);

                callback(result);
            }, function (tx, error) {
```

```
                   console.log("Got error in isFav error.code =" + error.code + "
error.message = " + error.message);
                   callback(false);


              });
          });
     } catch (err) {
         console.log("Got error in isFav " + err);
         callback(false);
     }


}

//-------------------------------------------------------------------------------
/**
 * Binding Search button handler to go and fetch place results
 */

function initiateSearch() {
    $("#search").click(function () {
        try {
            $.mobile.showPageLoadingMsg();


            navigator.geolocation.getCurrentPosition(function (position) {

                var radius = $("#range").val() * 1000;
                mapdata = new
                google.maps.LatLng(position.coords.latitude, position.coords.longitude);
                var url =
"https://maps.googleapis.com/maps/api/place/search/json?location=" +
position.coords.latitude + "," + position.coords.longitude + "&radius=" + radius +
"&name=" + $("#searchbox").val() + "&sensor=true&key=<API_Key>";
                $.getJSON(url, function (data) {
                    cachedData = data;
                    $("#result-
list").html("");
                    try {

                        $(data.results).each(function (index, entry) {


                            var htmlData = "<a href=\"#details\" id=\"" +
entry.reference + "\"><img
src=\"" + entry.icon + "\" class=\"ui-li-icon\"></img><h3> " + entry.name +
"</h3><p><strong> vicinity:" + entry.vicinity + "</strong></p></a>";

                            var liElem = $(document.createElement('li'));

                            $("#result-list").append(liElem.html(htmlData));

                            $(liElem).bind("tap", function (event) {

                                event.stopPropagation();
```

```
                            fetchDetails(entry);

                            return true;
                        });
                    });

                    $("#result-list").listview('refresh');
                } catch (err) {

                        console.log("Got error while putting search result on result
page " + err);
                    }
                    $.mobile.changePage("list");
                    $.mobile.hidePageLoadingMsg();
                }).error(function (xhr, textStatus, errorThrown) {
                    console.log("Got error while fetching search result : xhr.status=" +
xhr.status);

                }).complete(function (error) {
                    $.mobile.hidePageLoadingMsg();
                });
            }, function (error) {
                console.log("Got Error fetching geolocation " + error);
            });

        } catch (err) {
            console.log("Got error on clicking search button " + err);
        }
    });

}

//------------------------------------------------------------------


function bind() {
    initiateMap();
    initiateFavorites();
    initiateSearch();
    initiateFavButton();
}
//---------------------------------------------------

function onDeviceReady() {
    $(document).ready(function () {
        bind();
    });
}
document.addEventListener("deviceready", onDeviceReady);
//------------------------
```

The complete source of the app.css is as follows

```
#map, .map-content, #map_canvas {
    width: 100%;
    height: 100%;
```

```
    padding: 0;
}

#map_canvas {
    height: min-height: 100%;
}
```

Pros of jQueryMobile

jQueryMobile is an easy to use JavaScript UI framework for mobile application developers. The best part of jQueryMobile is its declarative UI programming. Using HTML tags to create a UI and adding "data-Role" pages to annotate the HTML tags as a page, header, footer, content, list, and button make it very easy to quickly program the UI layout.

Another great thing about jQueryMobile is the idea of pages. Pages are declared as divs in an HTML page. Also, the navigation part and history management part are built into jQueryMobile. This takes the headache out of putting in the history management part.

While jQueryMobile provides good support for widgets and toolbars, the programming aspect of these requires that the developer do DOM manipulation.

The biggest strength of jQueryMobile is that it is built on the robust jQuery core framework. Another strength of jQueryMobile is that it supports iOS, Android, BlackBerry, HP WebOS, Nokia/Symbian, Windows Mobile, Opera Mobile/Mini, Firefox mobile, and all modern desktop browsers.

jQueryMobile provides touch events for mobile and tablet applications. jQueryMobile also provides great theme support. Switching between themes is as simple as changing an attribute on an HTML tag.

Cons of jQueryMobile

jQueryMobile is a great lightweight framework and using jQuery to manipulate the DOM allows a user to build applications in a nice and easy manner. However as the complexity of an application increases, and the need for data model and corresponding views arrive, programming in jQueryMobile is more of implementing your own MVC framework in JavaScript.

In short, jQueryMobile is difficult to use when your application is complex. The lack of MVC framework or even Models and JavaScript Views makes JavaScript UI programming in jQueryMobile painful in comparison to other frameworks like Sencha Touch.

Conclusion

jQueryMobile is a great JavaScript mobile UI development framework if the mobile applications are fairly simple. As the complexity of your mobile application UI increases, programming in jQueryMobile will become more cumbersome.

Chapter 5

Using PhoneGap with Sencha Touch

Sencha Touch is a product from a company originally called "ExtJS." "ExtJS" is a popular company in the Ajax RIA world that provides a rich polished JavaScriptui framework named "ExtJS". The popular products of this company are "ExtJS" JavaScriptui framework, "Ext-GWT", GWT UI framework (GWT counter part of ExtJS), and "Sencha Touch" JavaScript library for mobiles.

"ExtJS" company was recently renamed to "Sencha". So, while the name is new, what goes in "Sencha Touch" library is based on years of experience building UI using JavaScript.

If you are familiar with "ExtJS", you will notice many similarities between "ExtJS" and "Sencha Touch", especially in the foundation classes. However, "Sencha Touch" is designed and intended only for mobile applications.

Why Use Sencha Touch?

Sencha Touch allows you to develop browser-based applications for iPhone, Android, and BlackBerry with a native look and feel. Also, Sencha Touch is based on HTML5.

Sencha Touch provides you with the following advantages:

1. Touch-optimized rich widget set, as well as touch-events that support tap, double tap, swipe, tap and hold, pinch and rotate, slide and gestures.

2. New age web standards HTML5 and CSS3.

3. Integrates with PhoneGap.

4. Supports iOS, Android, and BlackBerry, along with native themes for these devices

5. Out of box support for Ajax, JSONP, and Yahoo! Query Language (YQL), as well as support for local storage to back the widgets.

In short, Sencha Touch is currently one of the best JavaScript ui libraries for mobile application development. The learning curve may be a little steep at first, if you haven't worked on ExtJS.

Pros of Sencha Touch

The pros of Sencha Touch far exceed the cons. First of all, Sencha Touch is based on web standards, like HTML5 and CSS3, and not on any proprietary technology. The community support of Sencha Touch is also very good. It's free for commercial use.

You can build an application in Sencha Touch that can detect whether we are on a tablet or a phone and you can code to make the same application work differently. For example, look at the kitchen sink. When you view it on a tablet and on a mobile, you will see the example adjusts itself to make use of the real estate.

The widget set is quite rich. Building the entire widget in JavaScript ensures a high degree of interactivity with the user. You have more control.

The performance of Sencha Touch is good and improving with every release. Also, more with newer versions of iOS and Android OS releases, the webkit shipped with these OS is also improving in performance.

There is support for internationalization and there are widgets like grids and carousels that are very new age visualization aids.

Cons of Sencha Touch

The biggest con of Sencha Touch is its learning curve. With Sencha Touch, you rarely use any pre-rendered HTML. Everything is added to the DOM through JavaScript. It might be a conceptual shift for some people.

Sencha Touch would be overkill if your application is simply a few pages with navigation and the views are mostly list views, forms, and toolbars.

Downloading Sencha Touch

Download Sencha Touch library from Sencha's web site–
`www.sencha.com/products/touch/`. Once you download and unzip the sdk, you will see the structure shown in Figure 5–1.

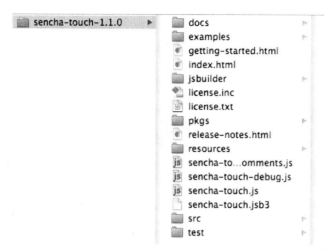

Figure 5–1. *Sencha Touch directory structure*

Integrating Sencha with PhoneGap

Let's begin with integrating Sencha Touch with a PhoneGap project. This chapter will assume it is for an Android platform. The steps for other platforms are similar.

Refer to Chapter 2 and Chapter 3 to setup your PhoneGap project for whichever platform you are aiming for.

As shown in Figure 5–2, from the Sencha Touch sdk, you will need to add the following files:

1. Add the sencha-touch.js JavaScript file to www/lib.

2. Add the resources/css folder to www/lib

3. Put all the application code in a file named app/app.js. This will be our main JavaScript file.

For the sake of this chapter and example, copy the icon.png, phone_startup.png, and tablet_startup.png from sencha-touch-1.1.0/examples/map folder.

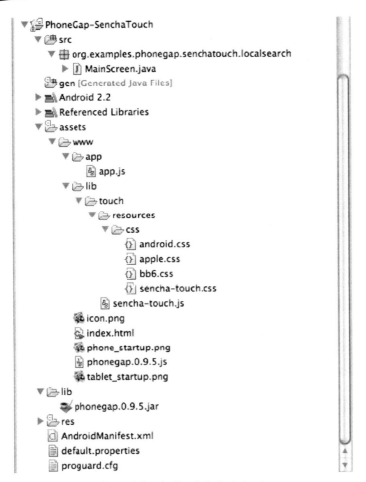

Figure 5–2. *PhoneGap and Sencha Touch Project structure*

Building a Local Search Application Using Sencha Touch

The requirements for the local search application are similar to what we have in Chapter 5. User enters a search by keyword, a range to search from his/her current location and user get local places listed in a list view. User can click on one of the items and see the details of the place. On the details screen, users can choose to put the place in his/her favourite list (stored in application database for offline access). The user can also see the search results in a map view.

Last but not least, users can click on the favourite button (star icon) to see his/her favourite listing.

Let's begin building the application. Remember Sencha Touch is fairly large and this chapter will run you through a subset of Sencha Touch's API required for this application.

Initializing Sencha Touch

The first step is to make sure index.html has Sencha Touch library, PhoneGap library, and CSS linked. Note our body is empty. This is because, in Sencha Touch, we build our entire ui in JavaScript. Note we are including the following JavaScript and stylesheet.

1. Sencha Touch stylesheet

2. Google Maps JavaScript

3. Our application JavaScript

```
<!DOCTYPE HTML>
<html>

    <head>
        <title>Local Search</title>
        <link rel="stylesheet" type="text/css" href="lib/touch/resources/css/sencha-
touch.css"></link>
        <!-- applications that determine the user's location via a sensor must
        passsensor=true when loading the Maps API JavaScript. -->
        <script type="text/javascript"
src="http://maps.google.com/maps/api/js?sensor=true"></script>
        <script type="text/javascript" src="lib/touch/sencha-touch.js"></script>
        <script type="text/javascript" src="phonegap-1.1.0.js"></script>
        <script type="text/javascript" src="app/app.js"></script>
    </head>

    <body>
    </body>

</html>
```

Now let's head over to app.js. The PhoneGap application is setup in a function Ext.setup(). As a rule of thumb, remember all Sencha Touch functions take a JSON structure as a configuration.

We will do the same to Ext.setup. See the code below. We will provide some icons for the application, as well as phone splash and tablet splash screens. But these are not the parts we want to focus on in this chapter.

The most important part here is onReadyfunction (). This is like jQuery's document ready and PhoneGap's device's ready function. We can begin drawing Sencha Touch's UI in this function.

```
Ext.setup({
    tabletStartupScreen: 'tablet_startup.png',
    phoneStartupScreen: 'phone_startup.png',
    icon: 'icon.png',
    glossOnIcon: false,
    onReady: function () {
```

```
            //Sencha Touch framework has initialized here.
            //Create Panels and bind event handlers.
    }
});
```

Creating the Layout (Application Skeleton)

The next step is to declare a panel as your main panel. To do this, we will create a new panel (new Ext.Panel()) and, in its configuration JSON, we will declare the following:

1. layout: 'card' - the layout is to be a card layout, which means it's a stack of cards, and we will show only one card at one time

2. fullscreen: true – specifies this panel will take up 100% width and height available and automatically draws itself to the page

3. items: [searchPanel,tabResultPanel,favourites, resultDetailPanel] – an array of child components to be added to this panel. As we are using a 'card' layout, it will show one child component at a time. There are four children in our main panel: searchPanel, tabResultPanel, favourites, and resultDetailPanel. searchPanel and tabResultPanel are declared in the same manner as mainPanel. By default, searchPanel is the visible card while other panels hide behind searchPanel

4. dockedItems: [] – use to declare docked widgets, typically for toolbar buttons.

5. Inside dockedItems has a toolbar declared in it by JSON representation. This toolbar has two buttons and one spacer to separate them.

    a. For the home button, we use the iconCls:'home'

    b. For the favourite button, we use the iconCls:'star'

For both these buttons, we have declared a handler, which is called when the buttons are clicked.

```
//Main Panel with CardLayout
var mainPanel = new Ext.Panel({
    layout: 'card',
    fullscreen: true,
    items: [searchPanel, tabResultPanel, favorites, resultDetailPanel],
    dockedItems: [{
        xtype: 'toolbar',
        title: 'Local Search',
        dock: 'top',
        items: [{
            iconMask: true,
            ui: 'round',
            iconCls: 'home',
            handler: function () {

            }

        }, {
```

```
            xtype: 'spacer'
        }, {
            iconMask: true,
            ui: 'round',
            iconCls: 'star',
            handler: function () {}

        }]
    }]
});
```

Without any of its child widgets, Figure 5–3 shows how this panel will appear.

Figure 5–3. *Main application panel with toolbar buttons*

Next, we declare the search panel. The search panel has a text box where users can enter the search keyword and has a range selector to allow the user to choose the range of his/her search. Finally, the search panel has a toolbar with a search button. This is declared as follows:

```
var searchPanel = new Ext.form.FormPanel({
    layout: 'fit',
    fullscreen: true,
    scroll: 'vertical',
    standardSubmit: false,
    //Adding form field
    items: [{
```

```
            xtype: 'fieldset',
            title: 'Local Search',
            items: [{
                xtype: 'textfield',
                name: 'search',
                label: 'Search',
                value: 'Pizza',
                userClearIcon: true,
                autoCapitalize: false
            }, {
                xtype: 'sliderfield',
                name: 'range',
                label: 'Range (0-10 Kms)',
                value: 5,
                minValue: 0,
                maxValue: 10
            }]

        }],
        //Docking a toolbar at bottom
        dockedItems: [{
            xtype: 'toolbar',
            dock: 'bottom',
            items: [{
                xtype: 'spacer'
            }, {
                text: 'Search',
                iconCls: 'search',
                title: 'Search',
                iconMask: true,
                ui: 'confirm',
                handler: function () {

                }
            }]
        }]
    });
```

Figure 5–4 shows how the search panel will appear.

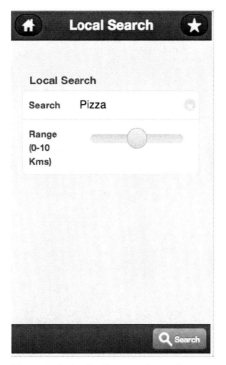

Figure 5–4. *Application search panel*

When the user does his/her search, the user is presented with two views.

1. List view showing the search result

2. Map view showing the search result

Both of these views are encapsulated inside a tab panel. We declare the tab panel as follows:

```
var tabResultPanel = new Ext.TabPanel({
    layout: 'fit',
    tabBar: {
        dock: 'bottom',
        layout: {
            pack: 'center'
        }
    },
    items: [result, map],

});
```

The tab panel without any of its child panels is shown in Figure 5–5. Note that we defined in the configuration JSON to position the tab bar on the bottom for the two tabs 'result' and 'map'. By default, the 'result' tab will be selected. We will define label and icon for both tabs when we create the 'result' and 'map' object.

Figure 5–5. *Search Results panel with tabs*

Now, we will see how to show the search result when a user does his/her searches. The part of how we do an AJAX call in the section is covered at a later part of this chapter.

For the sake of this chapter, assume you have a JSON coming from the Google place server, which looks as follows:

```
{
    "status": "OK",
    "results": [{
        "name": "Zaaffran Restaurant - BBQ and GRILL, Darling Harbour",
        "vicinity": "Darling Drive, Darling Harbour, Sydney",
        "types": ["restaurant", "food", "establishment"],
        "geometry": {
            "location": {
                "lat": -33.8712950,
                "lng": 151.1984770
            }
        },
        "icon": "http://maps.gstatic.com/mapfiles/place_api/icons/restaurant-71.png",
        "reference": "CpQBiwAAANM1CkdWcBxiExHinloJpp7kX2D3nyb_DOqoQ_-RuBhq9cwJKYvU8-
sRJUaXF4U2kET_OH3Oh3Yz4tf5_6gBgcsFAPyRappCrJ5WksvMkXrT5lA7q9U_SOZIOu3mrsvTtXnTDMKlBMywE_
5Yy6lbshqPIatWZ6QkPZBNdmkifyN3vM7H2vL-
300iY6EoartWuxIQNckbMOBs4D946thThmKOsBoUCmGgFrtYgtOOCIUc79fQi3waOOw",
        "id": "677679492a58049a7eae079e0890897eb953d79b"
    }, {
        "name": "Toros Restaurant Darling Harbour",
        "vicinity": "Murray Street, Sydney",
```

```
            "types": ["restaurant", "food", "establishment"],
            "geometry": {
                "location": {
                    "lat": -33.8714080,
                    "lng": 151.1975410
                }
            },
            "icon": "http://maps.gstatic.com/mapfiles/place_api/icons/restaurant-71.png",
            "reference": "CoQBdQAAALFujBuIMYXsG8Qlus2zSHeikZQNCsSbeIIo-55zkhCiArbPkACXRU-
CcLZbeKsXaBpoBNH5iyYJg6Nquct2LTE127X4CD1YtKpozmbjZpyCRFrJ_V5DI4IDGLCWeY_8NMxznbiqb9prR8m
XJoAKv7jNz6KEMxAuGLRAXbi7G6CYEhBeR6Ur-x2AB1S3pKXsKXLvGhRWFzL3Q5TOOxe-gm_LJm9cgtzYJw",
            "id": "aefbc59325ffd5f3e93d67932375d20d143289de"
        }, {
            "name": "Strike Bowling Bar Darling Harbour",
            "vicinity": "Sydney",
            "types": ["restaurant", "food", "establishment"],
            "geometry": {
                "location": {
                    "lat": -33.8662990,
                    "lng": 151.2016580
                }
            },
            "icon": "http://maps.gstatic.com/mapfiles/place_api/icons/restaurant-71.png",
            "reference": "CoQBeAAAAO-prCRp9Atcj_rvavsLyv-
DnxbGkw8QyRZb6Srm6QHOcww6lqFhIs2c7Ie6fMg3PZ4PhicfJL7ZWlaHaLDTqmRisoTQQUn61WTcSXAAiCOzcmO
JDBnafqrskSpFtNUgzGAOx29WGnWSP44jmjtioIsJN9ik8yjK7UxP4buAmMPVEhBXPiCfHXk1CQ6XRuQhpztsGhQ
U4U6-tWjTHcLSVzjbNxoiuihbaA",
            "id": "0a4e24c365f4bd70080f99bb80153c5ba3faced8"
        }
        ...additional results...],
        "html_attributions": ["Listings by \u003ca
href=\"http://www.yellowpages.com.au/\"\u003eYellow Pages\u003c/a\u003e"]
}
```

Now that we have seen the JSON structure of the Google places result, we will create
the panel to show the result. In this case, we are extending a component and declaring a
template (tpl) in the component.

Templating is a feature of Sencha Touch where you will declare an html in <tpl> tags. In
our case, we are passing the above JSON's results object. The results object is actually
an array. In our template code, notice the <tpl for=".">. This is telling the Sencha
template engine that iterates over all the objects inside the results.

In the later section of the html, you will notice placeholders like {reference}, {icon},
{name} etc. If any one of you has worked with java's message formatting, you will notice
this is quite similar. These {} entries will be replaced with corresponding data from the
JSON.

{name} will be replaced by the name inside results->entry->name.

To populate this panel with data, we will call the following API:

```
//This will call the template engine and draw the AJAX's response
    //result. Here 'result' is the Component object to show the results
    //HTML and response.results is the JSON array.
result.update(response.results);
```

Now, let's see the code that is used to create the result panel.

```
var result = new Ext.Component({

    title: 'Search Result',
    iconMask: true,
    iconCls: 'organize',
    cls: 'timeline',
    scroll: 'vertical',
    tpl: ['<tpl for=".">',
          '<div class="place" id="{reference}">',
          '<div class="icon"><imgsrc="{icon}" /></div>',
          '<div>', '<h2>{name}</h2>',
          '<p>{vicinity}</p>',
          '</div>',
          '</div>',
          '</tpl>'],
    listeners: {
        el: {
            tap: detailClickHandler,
            //function which
            //will handle tap event
        }
    }
});
```

Notice the listeners and el part at the end. This is telling Sencha Touch that we are interested to receive events on the elements of this component. Furthermore, we are telling it that we are specifically looking for tap events. The result of this code is that whenever the user taps on any of the places listed in the results, it will call the detailClickHandler function.

Figure 5–6. *Search Results panel*

The Map widget of Sencha Touch makes life much easier. Otherwise, we would have to work with Google APIs for maps. We simply create a new Ext.Map and give it some options. This is the simplest way to get a map going. Note that 'map' object will be used in AJAX callback to add the place markers on it. The AJAX call is described in 'Fetching the Places Listing'.

```
var map = new Ext.Map({
    iconMask: true,
    iconCls: 'maps',
    title: 'Map',
    // Name that appears on this tab
    mapOptions: {
        // Used in rendering map
        zoom: 12
    }
});
```

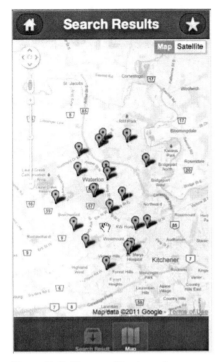

Figure 5–7. *Map panel to show places*

Next in line is the panel, which shows the details of a place. Note, once the user clicks on a search entry, the application will fetch the details from the Google places server. The JSON response of this request appears as follows:

```
{
    "status": "OK",
    "result": {
        "name": "Google Sydney",
        "vicinity": "Pirrama Road, Pyrmont",
        "types": ["establishment"],
        "formatted_phone_number": "(02) 9374 4000",
        "formatted_address": "5/48 Pirrama Road, Pyrmont NSW, Australia",
        "address_components": [{
            "long_name": "48",
            "short_name": "48",
            "types": ["street_number"]
        }, {
            "long_name": "Pirrama Road",
            "short_name": "Pirrama Road",
            "types": ["route"]
        }, {
            "long_name": "Pyrmont",
            "short_name": "Pyrmont",
            "types": ["locality", "political"]
        }, {
            "long_name": "NSW",
            "short_name": "NSW",
```

```
                "types": ["administrative_area_level_1", "political"]
            }, {
                "long_name": "2009",
                "short_name": "2009",
                "types": ["postal_code"]
            }],
            "geometry": {
                "location": {
                    "lat": -33.8669710,
                    "lng": 151.1958750
                }
            },
            "rating": 4.5,
            "url": "http://maps.google.com/maps/place?cid=10281119596374313554",
            "icon": "http://maps.gstatic.com/mapfiles/place_api/icons/generic_business-
71.png",
            "reference":
"CmRRAAAAUgylGnuntxKOuZy9_c5zxdFi6e491_FvOm1hks5YkeaH7k1SP9ujAkG4GROr1XCHFnMsDhuEIgQQq2W
Wyd33oGRAT8Vwr8rjTWEYEMvCZ1RxTzXSVDZ4gEFqLZcRyAw_EhBS8uZHidMMbYHuf9KHapRyGhQQ1dnf3uMghMR
BlXqJE6ygh_a3ag",
            "id": "4f89212bf76dde31f092cfc14d7506555d85b5c7"
        },
        "html_attributions": []
}
```

The idea is to display the above information to the user in a tabular manner. To do so, we will use a combination of the following:

1. A template to show the above JSON as a part of a table

2. A button that will allow users to add this place as a favourite or remove it as a favourite.

Hence, we use a wrapper panel called resultDetailPanel. This panel has a vbox layout (stacks widgets vertically). The first child is placeDetailsPanel (see below) and second is a button.

The button's text changes from "add to Favorite" to "Remove from Favorite", depending on whether the user has already made the place a favorite . There is a function defined in the application for the same that is named isFav().

```
var resultDetailPanel = new Ext.Panel({
    layout: {
        type: 'vbox',
    },
    items: [
    placeDetailsPanel,
    {
        xtype: 'button',
        text: 'Add to Favorite',
        handler: function (button, event) {
            if (button.text == "Add to Favorite") {
                addCurrentToFav();
                button.setText("Remove from Favorite");
            } else {
```

```
                        removeCurrentFromFav();
                        button.setText("Add to Favorite");
                    }

                }

        }],
        dockedItems: [{
            xtype: 'toolbar',
            dock: 'bottom',
            items: [{
                ui: 'round',
                text: 'Back',
                handler: function () {}

            }]
        }]

    });
```

This is the panel that shows the JSON result as a table. It uses a template for the same. A template is an html template, which has placeholders represented by {<<variable>>}. The template code is used and the placeholders are replaced by actual values:

```
var placeDetailsPanel = new Ext.Panel({
    tpl: ['<table>',
            '<tr>',
            '<td>',
            '</td>',
            '<td>',
            '<h1 class="bold">Business Details</h1>',
            '</td>',
            '</tr>',
            '<tr>',
            '<td>',
            '<h1 class="bold">Name</h1>',
            '</td>',
            '<td>',
            '<h1>{name}</h1>',
            '</td>',
            '</tr>',
            '<tr>',
            '<td>',
            '<h1 class="bold">Address</h1>',
            '</td>',
            '<td>',
            '<h1>{formatted_address}</h1>',
            '</td>',
            '</tr>',
            '<tr>',
            '<td>',
            '<h1 class="bold">Phone</h1>',
            '</td>',
```

```
'<td>',
'<h1>{formatted_phone_number}</h1>',
'</td>',
'</tr>',
'<tr>',
'<td>',
'<h1 class="bold">Rating</h1>',
'</td>',
'<td>',
'<h1>{rating}</h1>',
'</td>',
'</tr>',
'<tr>',
'<td>',
'<h1 class="bold">Home Page</h1>',
'</td>',
'<td>',
'<a href="{url}" target="_blank">Home Page</a>',
'</td>',
'</tr>',
'</table>'

    ]
});
```

The place details panel is shown in Figure 5–8.

Figure 5–8. *Place Details panel*

Now that we have seen how to add or remove places to and from favorites, let's see the panel, which lists user's favorite places. And yes, this panel is quite similar to the results panel. The only difference is the result panel is given the JSON, which comes from Google places server, and the favorites panel is given the JSON, which comes from the database.

```
var favorites = new Ext.Component({
    title: 'Favotites',
    iconMask: true,
    iconCls: 'organize',

    cls: 'timeline',
    scroll: 'vertical',
    tpl: ['<tpl for=".">',
            '<div class="place" id="{reference}">',
            '<div class="icon"><imgsrc="{icon}" /></div>',
            '<div>',
            '<h2>{name}</h2>',
            '<p>{vicinity}</p>',
            '</div>',
            '</div>',
            '</tpl>'],
    listeners: {
        el: {
            tap: detailClickHandler,
        }
    }
});
```

Figure 5–9. *Favorite places panel*

Switching Between Panels

As you explore more of the application code, you will need to switch from one panel to another panel. This is the main reason we are using a 'card' layout for the main panel.

In order to switch from one panel to another panel in card layout, we will use the following code. Note the first argument index number for widgets supplied in items to the mainPanel. The second argument is an animation effect.

```
mainPanel.setActiveItem(0, "slide");
mainPanel.setActiveItem(1, {type: 'slide', direction: 'right'});
```

Also, the main panel owns the toolbar. We need to change the title of this toolbar to let the user know where he is. This is done using the following code:

```
mainPanel.dockedItems.items[0].setTitle('Details');
```

Fetching the Places Listing

When a user clicks on the search button on the search panel, we need to make an Ajax call to fetch Google places' result. The following function shows how to do the same in Sencha Touch and PhoneGap.

The steps are quite simple.

1. Get the geo location from PhoneGap

2. In the success call of getcurrentPosition, initiate an Ajax call by calling Ext.ajax.request(url,successCallback,failureCallback)

3. In the successCallback of Ext.ajax.request, you will get the JSON string.

   a. Convert this JSON string to JSON object

   b. Populate the result panel by calling result.update(obj.results);

   c. Populate the market on the Google maps

```
var fetchFromGoogle = function () {

        var keyword = searchPanel.items.items[0].items.items[0].value;
        var range = searchPanel.items.items[0].items.items[1].value * 1000;
        navigator.geolocation.getCurrentPosition(

        function (position) {
            var lat - position.coords.latitude;
            var lng = position.coords.longitude;


            map.update({
                latitude: lat,
                longitude: lng
            });

            var googlePlaceUrl =
'https://maps.googleapis.com/maps/api/place/search/json?location='
                    + lat + ',' + lng + '&radius=' + range + '&types=food&name=' + keyword
+ '&sensor=true&key=API_Key';
                //Note that you will need to replace the API_Key with your own key. You
//can get API Key from
                //http://code.google.com/apis/maps/documentation/places/
                Ext.Ajax.request({
                    url: googlePlaceUrl,
                    success: function (response, opts) {


                        var obj = Ext.decode(response.responseText);

                        result.update(obj.results);
                        var data = obj.results;
                        for (var i = 0, ln = data.length; i < ln; i++) { // Loop to add
points to the map
                            var place = data[i];

                            if (place.geometry && place.geometry.location) {
                                var position = new
google.maps.LatLng(place.geometry.location.lat, place.geometry.location.lng);
                                addMarker(place.name, place.reference, position); // Call
addMarker function with new data
                            }
```

```
                    }
                },
                failure: function (response, opts) {
                    console.log('server-side failure with status code ' +
response.status);

                }
            }, function (err) {
                console.log('Failed to get geo location from phonegap ' + err);
            });
        })
    }
```

Fetching Places Details

Fetching the places details from the Google places server is even easier. You need even fewer things.

1. Initiate an Ajax call by calling Ext.ajax.request(url,successCallback,failureCallback)

2. In the successCallback of Ext.ajax.request, you will get the JSON string.

    a. Convert this JSON string to a JSON object

    b. Populate the result panel by calling placeDetailsPanel.update(obj.result);

    c. Also we will check whether this place is favorite by calling isFav() function

        i. If the place is a favorite, we rename the button to "remove from Favorite"
        ii. Otherwise, we rename the button to "add to Favorite"

```
var cachedDetails = null;

/**
 * Ensure we have the table before we use it
 * @param {Object} tx
 */
var ensureTableExists = function (tx) {
    tx.executeSql('CREATE TABLE IF NOT EXISTS Favourite (id unique, reference,
                        name,address,phone,rating,icon,vicinity)');
}

/**
 * Add currentDetails to DB
 */
var addCurrentToFav = function () {
    addToFavorite(cachedDetails);
}

/**
```

```
     * Remove currentDetails from DB
     */
    var removeCurrentFromFav = function () {
        removeFromFavorite(cachedDetails);
    }


    /**
     * Add current business data to favourite
     * @param {Object} data
     */
    var addToFavorite = function (data) {
        var db = window.openDatabase("Favourites", "1.0", "Favourites", 20000000);

        db.transaction(function (tx) {

            ensureTableExists(tx);

            var id = (data.id != null) ? ('"' + data.id + '"') : ('""');
            var reference = (data.reference != null) ? ('"' + data.reference + '"') :
    ('""');
            var name = (data.name != null) ? ('"' + data.name + '"') : ('""');
            var address = (data.formatted_address != null) ? ('"' + data.formatted_address +
    '"') : ('""');
            var phone = (data.formatted_phone_number != null) ? ('"' +
    data.formatted_phone_number + '"') : ('""');
            var rating = (data.rating != null) ? ('"' + data.rating + '"') : ('""');
            var icon = (data.icon != null) ? ('"' + data.icon + '"') : ('""');
            var vicinity = (data.vicinity != null) ? ('"' + data.vicinity + '"') : ('""');


            var insertStmt = 'INSERT INTO Favourite (id,reference,
                            name,address,phone,rating,icon,vicinity) VALUES
                            (' + id + ',' + reference + ',' + name + ',' + address + ','
                            + phone + ',' + rating + ',' + icon + ',' + vicinity + ')';


            tx.executeSql(insertStmt);

        }, function (error) {
            console.log("Data insert failed " + error.code + "   " + error.message);
        }, function () {
            console.log("Data insert successful");
        });

    }


    /**
     * Remove current business data from favourite
     * @param {Object} data
     */
    var removeFromFavorite = function (data) {
        try {
            var db = window.openDatabase("Favourites", "1.0", "Favourites", 20000000);
```

```
        db.transaction(function (tx) {
            ensureTableExists(tx);
            var deleteStmt = "DELETE FROM Favourite WHERE id = '" + data.id + "'";
            console.log(deleteStmt);
            tx.executeSql(deleteStmt);

        }, function (error) {
            console.log("Data Delete failed " + error.code + "    " + error.message);
        }, function () {
            console.log("Data Delete successful");
        });

    } catch (err) {
        console.log("Caught exception while deleting favourite " + data.name);
    }

}

/**
 *
 * @param {Object} reference
 * @return true if place is favourite else false
 */
var isFav = function (data, callback) {

    var db = window.openDatabase("Favourites", "1.0", "Favourites", 200000);

    try {
        db.transaction(function (tx) {
            ensureTableExists(tx);

            var sql = "SELECT * FROM Favourite where id='" + data.id + "'";

            tx.executeSql(sql, [], function (tx, results) {
                var result = (results != null && results.rows != null &&
results.rows.length > 0);
                callback(result);
            }, function (tx, error) {
                var fetchDetails = function (reference) {
                    placeDetailsPanel.update({
                        name: "",
                        formatted_address: "",
                        formatted_phone_number: "",
                        rating: "",
                        url: ""
                    });
                    Ext.Ajax.request({
                        url:
'https://maps.googleapis.com/maps/api/place/details/json?reference='
                            + reference + '&sensor=true&key=API_Key',
                        success: function (response, opts) {
                            var obj = Ext.decode(response.responseText);
                            //global variable to store the current place
```

```
                                    cachedDetails = obj.result;
                                    isFav(obj.result, function (result) {
                                        if (result) {

                                            resultDetailPanel.items.items[1].setText("Remove
from Favorite");
                                        } else {

                                            resultDetailPanel.items.items[1].setText("Add to
Favorite");
                                        }
                                        placeDetailsPanel.update(obj.result);
                                    });

                                },
                                failure: function (response, opts) {
                                    console.log('server-side failure with status code ' +
response.status);
                                }
                            })
                        }

                    console.log("Got error in isFaverror.code =" + error.code + "
                        error.message = " + error.message);
                    callback(false);
                });
            });

        } catch (err) {
            console.log("Got error in isFav " + err);
            callback(false);
        }
    }
}
```

Storing and Retrieving Favorites from Database

The last important part of this application is to have functions defined to do the following:

1. Add a place to the favorite table

2. Remove a place from the favorite table

3. Check if a place is already in the favorite table

4. Get all entries from the favorite table.

In order to pass the place entry across various function calls, what we have done in this application is to declare a variable named cachedDetails. When we are in the page where we show the details of the place, we cache the current place in cachedDetails. cachedDetails is used to add a place to favorite, remove it from favorite, and also to check whether it is already a part of a user's favorites.

```
var cachedDetails = null;

/**
 * Ensure we have the table before we use it
 * @param {Object} tx
 */
var ensureTableExists = function (tx) {
    tx.executeSql('CREATE TABLE IF NOT EXISTS Favourite (id unique, reference,
                        name,address,phone,rating,icon,vicinity)');
}

/**
 * Add currentDetails to DB
 */
var addCurrentToFav = function () {
    addToFavorite(cachedDetails);
}

/**
 * Remove currentDetails from DB
 */
var removeCurrentFromFav = function () {
    removeFromFavorite(cachedDetails);
}

/**
 * Add current business data to favourite
 * @param {Object} data
 */
var addToFavorite = function (data) {
    var db = window.openDatabase("Favourites", "1.0", "Favourites", 20000000);

    db.transaction(function (tx) {

        ensureTableExists(tx);

        var id = (data.id != null) ? ('"' + data.id + '"') : ('""');
        var reference = (data.reference != null) ? ('"' + data.reference + '"') :
('""');
        var name = (data.name != null) ? ('"' + data.name + '"') : ('""');
        var address = (data.formatted_address != null) ? ('"' + data.formatted_address +
'"') : ('""');
        var phone = (data.formatted_phone_number != null) ? ('"' +
data.formatted_phone_number + '"') : ('""');
        var rating = (data.rating != null) ? ('"' + data.rating + '"') : ('""');
        var icon = (data.icon != null) ? ('"' + data.icon + '"') : ('""');
        var vicinity = (data.vicinity != null) ? ('"' + data.vicinity + '"') : ('""');


        var insertStmt = 'INSERT INTO Favourite (id,reference,
                        name,address,phone,rating,icon,vicinity) VALUES
                        (' + id + ',' + reference + ',' + name + ',' + address + ',' +
phone
                        + ',' + rating + ',' + icon + ',' + vicinity + ')';
```

```
        tx.executeSql(insertStmt);

    }, function (error) {
        console.log("Data insert failed " + error.code + "    " + error.message);
    }, function () {
        console.log("Data insert successful");
    });

}


/**
 * Remove current business data from favourite
 * @param {Object} data
 */
var removeFromFavorite = function (data) {
    try {
        var db = window.openDatabase("Favourites", "1.0", "Favourites", 20000000);

        db.transaction(function (tx) {
            ensureTableExists(tx);
            var deleteStmt = "DELETE FROM Favourite WHERE id = '" + data.id + "'";
            console.log(deleteStmt);
            tx.executeSql(deleteStmt);

        }, function (error) {
            console.log("Data Delete failed " + error.code + "    " + error.message);
        }, function () {
            console.log("Data Delete successful");
        });

    } catch (err) {
        console.log("Caught exception while deleting favourite " + data.name);
    }

}

/**
 *
 * @param {Object} reference
 * @return true if place is favourite else false
 */
var isFav = function (data, callback) {

    var db = window.openDatabase("Favourites", "1.0", "Favourites", 200000);

    try {
        db.transaction(function (tx) {
            ensureTableExists(tx);

            var sql = "SELECT * FROM Favourite where id='" + data.id + "'";
```

```
            tx.executeSql(sql, [],
                function (tx, results) {
                    var result = (results != null && results.rows != null &&
results.rows.length > 0);
                    callback(result);
                },
                function (tx, error) {
                    console.log("Got error in isFaverror.code =" + error.code + "
                        error.message = " + error.message);
                    callback(false);
                });
            });

    } catch (err) {
        console.log("Got error in isFav " + err);
        callback(false);
    }
}
```

This covers the layout part of the application, how to populate data from Google places server, and how to navigate and use database.

Please refer to the complete example listed below in order to learn how events are handled.

1. index.html

```
<!DOCTYPE HTML>
<html>
    <head>
        <title>Sencha Touch Layout</title>
        <link rel="stylesheet" type="text/css" href="lib/touch/resources/css/sencha-
touch.css"></link>
        <script type="text/javascript"
src="http://maps.google.com/maps/api/js?sensor=true"></script>
        <script type="text/javascript" src="lib/touch/sencha-touch.js"></script>
        <script type="text/javascript" src="app/app.js"></script>
        <style>
            .x-tabbar{
                padding-top: 10px;!important;
                border-bottom: 2px solid #306aa1 !important;
            }
            .place {
                padding: 10px 0 10px 68px;
                border-top: 1px solid #ccc;
                min-height: 68px;
                background-color: #fff;
            }
            .place h2 {
                font-weight:bold;
            }
            .place .icon {
                position: absolute;
                left: 10px;
            }
```

```
                .place .icon img{
                    height:24px;
                    width: 24px;
                }
                .bold{
                    font-weight: bold;
                }
            </style>
        </head>
        <body>
        </body>
    </html>
```

2. app.js

```
Ext.setup({
    tabletStartupScreen: 'tablet_startup.png',
    phoneStartupScreen: 'phone_startup.png',
    icon: 'icon.png',
    glossOnIcon: false,
    onReady: function () {
        var lastPanelId = 0;

        var SEARCHPAGE = 0;
        var TABPAGE = 1;
        var FAVPAGE = 2;
        var DETAILSPAGE = 3;

        var cachedDetails = null;

        var searchPanel = newExt.form.FormPanel({
            layout: 'fit',
            fullscreen: true,
            scroll: 'vertical',
            standardSubmit: false,
            //Adding form field
            items: [{
                xtype: 'fieldset',
                title: 'Local Search',
                items: [{
                    xtype: 'textfield',
                    name: 'search',
                    label: 'Search',
                    value: 'Pizza',
                    useClearIcon: true,
                    autoCapitalize: false
                }, {
                    xtype: 'sliderfield',
                    name: 'range',
                    label: 'Range (0-10 Kms)',
                    value: 5,
                    minValue: 0,
                    maxValue: 10
                }]
            }] //Docking a toolbar at bottom
            ,
```

```
        dockedItems: [{
            xtype: 'toolbar',
            dock: 'bottom',
            items: [{
                xtype: 'spacer'
            }, {
                text: 'Search',
                iconCls: 'search',
                title: 'Search',
                iconMask: true,
                ui: 'round',
                ui: 'confirm',
                handler: function () {
                    lastPanelId = TABPAGE;
                    fetchFromGoogle();

                    mainPanel.dockedItems.items[0].setTitle('Search Results');
                    mainPanel.setActiveItem(lastPanelId);
                }
            }]
        }]
});

var detailClickHandler = function (event) {
        var reference = event.getTarget(".place").id;
        fetchDetails(reference);
        mainPanel.dockedItems.items[0].setTitle('Details');
        mainPanel.setActiveItem(DETAILSPAGE, "slide");
    }

var result = new Ext.Component({

    title: 'Search Result',
    iconMask: true,
    iconCls: 'organize',
    cls: 'timeline',
    scroll: 'vertical',
    tpl: ['<tpl for=".">',
            '<div class="place" id="{reference}">',
            '<div class="icon"><imgsrc="{icon}" /></div>',
            '<div>',
            '<h2>{name}</h2>',

    '<p>{vicinity}</p>', '</div>', '</div>', '</tpl>'

    ],
    listeners: {
        el: {
            tap: detailClickHandler,
            delegate: '.place'

        }
    }

});

var favorites = new Ext.Component({
```

```
                    title: 'Favotites',
                    iconMask: true,
                    iconCls: 'organize',

                    cls: 'timeline',
                    scroll: 'vertical',
                    tpl: ['<tpl for=".">',
                            '<div class="place" id="{reference}">',
                            '<div class="icon"><imgsrc="{icon}" /></div>',
                            '<div>',
                            '<h2>{name}</h2>',
                            '<p>{vicinity}</p>',
                            '</div>',
                            '</div>',
                            '</tpl>'],
                    listeners: {
                        el: {
                            tap: detailClickHandler,
                            delegate: '.place'

                        }
                    }

        });

        var map = new Ext.Map({
            iconMask: true,
            iconCls: 'maps',
            title: 'Map',
            // Name that appears on this tab
            fullscreen: true,
            mapOptions: { // Used in rendering map
                zoom: 12
            }
        });

        var tabResultPanel = new Ext.TabPanel({
            layout: 'fit',
            tabBar: {
                dock: 'bottom',
                layout: {
                    pack: 'center'
                }
            },
            items: [result, map],


        });

        var placeDetailsPanel = new Ext.Panel({
            //layout: 'fit',
            tpl: ['<table>',
                    '<tr>',
                    '<td>',
                    '</td>',
                    '<td>',
                    '<h1 class="bold">Business Details</h1>',
```

```
                    '</td>',
                    '</tr>',
                    '<tr>',
                    '<td>',
                    '<h1 class="bold">Name</h1>',
                    '</td>',
                    '<td>',
                    '<h1>{name}</h1>',
                    '</td>',
                    '</tr>',
                    '<tr>',
                    '<td>',
                    '<h1 class="bold">Address</h1>',
                    '</td>',
                    '<td>',
                    '<h1>{formatted_address}</h1>',
                    '</td>',
                    '</tr>',
                    '<tr>',
                    '<td>',
                    '<h1 class="bold">Phone</h1>',
                    '</td>',
                    '<td>',
                    '<h1>{formatted_phone_number}</h1>',
                    '</td>',
                    '</tr>',
                    '<tr>',
                    '<td>',
                    '<h1 class="bold">Rating</h1>',
                    '</td>',
                    '<td>',
                    '<h1>{rating}</h1>',
                    '</td>',
                    '</tr>',
                    '<tr>',
                    '<td>',
                    '<h1 class="bold">Home Page</h1>',
                    '</td>',
                    '<td>',
                    '<a href="{url}" target="_blank">Home Page</a>',
                    '</td>',
                    '</tr>',
                    '</table>'

    ]
});

var resultDetailPanel = new Ext.Panel({
    layout: {
        type: 'vbox',
    },
    items: [
    placeDetailsPanel,
    {
        xtype: 'button',
        text: 'Add to Favorite',
        handler: function (button, event) {
```

```
                    if (button.text == "Add to Favorite") {
                        addCurrentToFav();
                        button.setText("Remove from Favorite");
                    } else {
                        removeCurrentFromFav();
                        button.setText("Add to Favorite");
                    }

                }

            }],
            dockedItems: [{
                xtype: 'toolbar',
                dock: 'bottom',
                items: [{
                    ui: 'round',
                    text: 'Back',
                    handler: function () {

                        if (lastPanelId == 0) {
                            mainPanel.dockedItems.items[0].setTitle('Home Page');
                        } else if (lastPanelId == 1) {
                            mainPanel.dockedItems.items[0].setTitle('Search Results');
                        } else if (lastPanelId == 2) {
                            fetchFromDB();
                            mainPanel.dockedItems.items[0].setTitle('Favourites');
                        } else if (lastPanelId == 3) {
                            //Shouldn't happen
                            mainPanel.dockedItems.items[0].setTitle('Details');
                        }

                        mainPanel.setActiveItem(lastPanelId, {
                            type: 'slide',
                            direction: 'right'
                        });
                    }

                }]
            }]

        });

        //Main Panel with CardLayout
        var mainPanel = new Ext.Panel({
            layout: 'card',
            fullscreen: true,
            items: [searchPanel, tabResultPanel, favorites, resultDetailPanel],
            dockedItems: [{
                xtype: 'toolbar',
                title: 'Local Search',
                dock: 'top',
                items: [{

                    iconMask: true,
                    ui: 'round',
                    iconCls: 'home',
                    handler: function () {
```

```
                    lastPanelId = SEARCHPAGE;

                    mainPanel.dockedItems.items[0].setTitle('Home Page');
                    mainPanel.setActiveItem(lastPanelId, "slide");
                }
            }, {
                xtype: 'spacer'
            }, {

                iconMask: true,
                ui: 'round',
                iconCls: 'star',
                handler: function () {
                    fetchFromDB();
                    lastPanelId = FAVPAGE;
                    mainPanel.dockedItems.items[0].setTitle('Favourites');
                    mainPanel.setActiveItem(lastPanelId, "slide");
                }

            }]
        }]
});

// These are all Google Maps APIs
var addMarker = function (name, reference, position) {

        var marker = new google.maps.Marker({
            map: map.map,
            position: position,
            clickable: true,
            optimized: true,
            title: name
        });
        google.maps.event.addListener(marker, 'click', function () {
            fetchDetails(reference);

            mainPanel.dockedItems.items[0].setTitle('Details');
            mainPanel.setActiveItem(DETAILSPAGE, "slide");

        });


    };



var fetchFromGoogle = function () {

        var keyword = searchPanel.items.items[0].items.items[0].value;
        var range = searchPanel.items.items[0].items.items[1].value * 1000;
        navigator.geolocation.getCurrentPosition(

        function (position) {
            var lat = position.coords.latitude;
            var lng = position.coords.longitude;
```

```
                    map.update({
                        latitude: lat,
                        longitude: lng
                    });

                    var googlePlaceUrl =
'https://maps.googleapis.com/maps/api/place/search/json?location='
                            + lat + ',' + lng + '&radius=' + range + '&types=food&name=' +
keyword + '&sensor=true&key=API_Key';
                        //Note that you will need to replace the API_Key with your own key.
You
                    //can get API Key from
//http://code.google.com/apis/maps/documentation/places/
                    Ext.Ajax.request({
                        url: googlePlaceUrl,
                        success: function (response, opts) {


                            var obj = Ext.decode(response.responseText);

                            result.update(obj.results);
                            var data = obj.results;
                            for (var i = 0, ln = data.length; i < ln; i++) { // Loop to
add points to the map
                                var place = data[i];

                                if (place.geometry && place.geometry.location) {
                                    var position = new
google.maps.LatLng(place.geometry.location.lat, place.geometry.location.lng);

                                    addMarker(place.name, place.reference, position); //
Call addMarker function with new data
                                }
                            }

                        },
                        failure: function (response, opts) {
                            console.log('server-side failure with status code ' +
response.status);

                        }
                    }, function (err) {
                        console.log('Failed to get geo location from phonegap ' + err);
                    });
                })
            }

        var fetchFromDB = function () {
                var db = window.openDatabase("Favourites", "1.0", "Favourites", 200000);
                try {
                    db.transaction(function (tx) {
                        tx.executeSql('SELECT * FROM Favourite', [], function (tx,
results) {
                            var arr = [];
                            for (var i = 0; i < results.rows.length; i++) {
```

```
                                    var data = results.rows.item(i)
                                    arr[i] = data;

                                }

                                favorites.update(arr);

                        }, function (error) {
                                console.log("Got error fetching favourites " + error.code +
" " + error.message);
                        });
                    });
                } catch (err) {
                    console.log("Got error while reading favourites " + err);
                }

            }



        var fetchDetails = function (reference) {
                placeDetailsPanel.update({
                    name: "",
                    formatted_address: "",
                    formatted_phone_number: "",
                    rating: "",
                    url: ""
                });
                Ext.Ajax.request({
                    url:
'https://maps.googleapis.com/maps/api/place/details/json?reference=' + reference +
'&sensor=true&key=API_Key',
                    success: function (response, opts) {
                        var obj = Ext.decode(response.responseText);
                        cachedDetails = obj.result;
                        isFav(obj.result, function (result) {
                            if (result) {

                                resultDetailPanel.items.items[1].setText("Remove from
Favorite");
                            } else {

                                resultDetailPanel.items.items[1].setText("Add to
Favorite");
                            }
                            placeDetailsPanel.update(obj.result);
                        });

                    },
                    failure: function (response, opts) {
                        console.log('server-side failure with status code ' +
response.status);
                    }
                })
            }
```

```javascript
        /**
         * Ensure we have the table before we use it
         * @param {Object} tx
         */
    var ensureTableExists = function (tx) {
            tx.executeSql('CREATE TABLE IF NOT EXISTS Favourite (id unique,
reference, name,address,phone,rating,icon,vicinity)');
            }

    var addCurrentToFav = function () {
            addToFavorite(cachedDetails);
            }

    var removeCurrentFromFav = function () {
            removeFromFavorite(cachedDetails);
            }

        /**
         * Add current business data to favourite
         * @param {Object} data
         */
    var addToFavorite = function (data) {
            var db = window.openDatabase("Favourites", "1.0", "Favourites",
20000000);

            db.transaction(function (tx) {
                ensureTableExists(tx);
                var id = (data.id != null) ? ('"' + data.id + '"') : ('""');
                var reference = (data.reference != null) ? ('"' + data.reference +
'"') : ('""');
                var name = (data.name != null) ? ('"' + data.name + '"') : ('""');
                var address = (data.formatted_address != null) ? ('"' +
data.formatted_address + '"') : ('""');
                var phone = (data.formatted_phone_number != null) ? ('"' +
data.formatted_phone_number + '"') : ('""');
                var rating = (data.rating != null) ? ('"' + data.rating + '"') :
('""');
                var icon = (data.icon != null) ? ('"' + data.icon + '"') : ('""');
                var vicinity = (data.vicinity != null) ? ('"' + data.vicinity + '"')
: ('""');
                var insertStmt = 'INSERT INTO Favourite (id,reference,
name,address,phone,rating,icon,vicinity) VALUES (' + id
                    + ',' + reference + ',' + name + ',' + address + ',' + phone +
',' + rating + ',' + icon + ',' + vicinity + ')';
                tx.executeSql(insertStmt);

            }, function (error) {
                console.log("Data insert failed " + error.code + "    " +
error.message);

            }, function () {
                console.log("Data insert successful");

            });

            }
```

```
        /**
         * Remove current business data from favourite
         * @param {Object} data
         */
      var removeFromFavorite = function (data) {
            try {
                var db = window.openDatabase("Favourites", "1.0", "Favourites",
20000000);

                db.transaction(function (tx) {
                    ensureTableExists(tx);
                    var deleteStmt = "DELETE FROM Favourite WHERE id = '" + data.id
+ "'";

                    console.log(deleteStmt);
                    tx.executeSql(deleteStmt);

                }, function (error) {
                    console.log("Data Delete failed " + error.code + "    " +
error.message);
                }, function () {
                    console.log("Data Delete successful");
                });
            } catch (err) {
                console.log("Caught exception while deleting favourite " +
data.name);
            }

        }

    /**
     *
     * @param {Object} reference
     * @return true if place is favourite else false
     */
    var isFav = function (data, callback) {

            var db = window.openDatabase("Favourites", "1.0", "Favourites", 200000);

            try {
                db.transaction(function (tx) {
                    ensureTableExists(tx);

                    var sql = "SELECT * FROM Favourite where id='" + data.id + "'";

                    tx.executeSql(sql, [], function (tx, results) {
                        var result = (results != null && results.rows != null &&
results.rows.length > 0);

                        callback(result);
                    }, function (tx, error) {
                        console.log("Got error in isFaverror.code =" + error.code +
" error.message = " + error.message);
                        callback(false);
```

```
            });
          });

        } catch (err) {
            console.log("Got error in isFav " + err);
            callback(false);



        }



      }


    }
});
```

Conclusion

If you are building a fairly complex mobile application, you should use Sencha Touch. jQueryMobile is good for smaller, less complex Ajax applications. Although jQueryMobile can be used for more complex applications, you will have to do DOM manipulation all by yourself and things will get more complex there.

Sencha Touch has a good performance and rich widget set. Some of its widgets use a data store to talk to the server components. You can use mvc design pattern with Sencha Touch and even split up your application into several .js files for improved modular code.

Chapter 6

Using PhoneGap with GWT

The Google Web toolkit (GWT) is a framework from Google that you can use to develop browser-based applications. The GWT allows developers to code in Java and generate JavaScript-based applications.

GWT applications are inherently cross-browser compatible and they are the smallest and fastest browser-based applications.

This chapter will focus on how to develop a GWT application for a mobile phone, using PhoneGap. The steps are based on the GWT PhoneGap library developed by Daniel Kurka. You can download this library from `http://code.google.com/p/gwt-phonegap/`.

Knowledge of how to develop GWT-based applications is essential. If you are new to GWT-based applications, you can visit `http://code.google.com/webtoolkit/doc/latest/tutorial/index.html` to learn more about GWT.

Why Use GWT for User Interface Development?

Before we jump into how to use GWT and PhoneGap together, let's first consider why GWT is a great choice for user interface development:

- The GWT allows developers to write browser-based applications without having to worry about cross browser issues, memory leaks in JavaScript, and the JavaScript language itself.

- The GWT allows developers to code in Java and compiles the user interface and business logic written in Java into JavaScript.

- The GWT also allows you to use concepts such as *deferred binding* (it's like runtime polymorphism for the JavaScript world). This approach allows developers to create a single application that can serve a mobile browser using different classes, and serve a desktop browser using another set of classes.

- The GWT ensures that you can create the smallest and fastest JavaScripts.

- The GWT is a well-accepted technology used by many companies and by a large segment of the developer community. The GWT is becoming the de facto choice for large, complex Ajax-based applications.

Along with the preceding advantages, many lightweight, ready-made widgets are available out of the box for the GWT. Also, there are professional GWT libraries, like EXT-GWT and Smart-GWT, which make the user interface look very professional and finished. Imagine your Java developers using their existing Java skills, writing browser-based applications with ease and with the best design practices. The GWT takes the pain of writing browser-based applications away, while delivering the best browser applications possible.

Getting Acquainted with the GWT PhoneGap

The GWT provides a mechanism named JavaScript Native Interface (JSNI), which allows the GWT to wrap over existing JavaScript libraries. This ability allows the developer to code in Java without having to worry about how the underlying JavaScript functions are invoked.

The GWT PhoneGap is a GWT wrapper for the PhoneGap library. The following section will demonstrate how to write a helloworld application using GWT PhoneGap.

Building a PhoneGap GWT Application

There are two main steps to building a GWT PhoneGap application. The first step is to build a GWT project. Once you have built the GWT project, the developer would compile the GWT project to create a web application (a set of HTMLs and JavaScripts).

The second step is to build an Android PhoneGap application (using version 0.9.4 of PhoneGap) and to embed the GWT web application into the PhoneGap application.

Build the GWT Application

You will need the following tools before you build a GWT application:

- JDK 1.6+
- Eclipse 3.6 Helios
- Eclipse Google Plugin
- PhoneGap 0.9.4 library
- GWT PhoneGap 0.8 version library
- Chrome browser 12+ version for testing

Create a new web application project (Google web application) and fill the wizard with the values shown in Figure 6–1. You will need to check "Use Google Web Toolkit" and uncheck "Use Google App Engine."

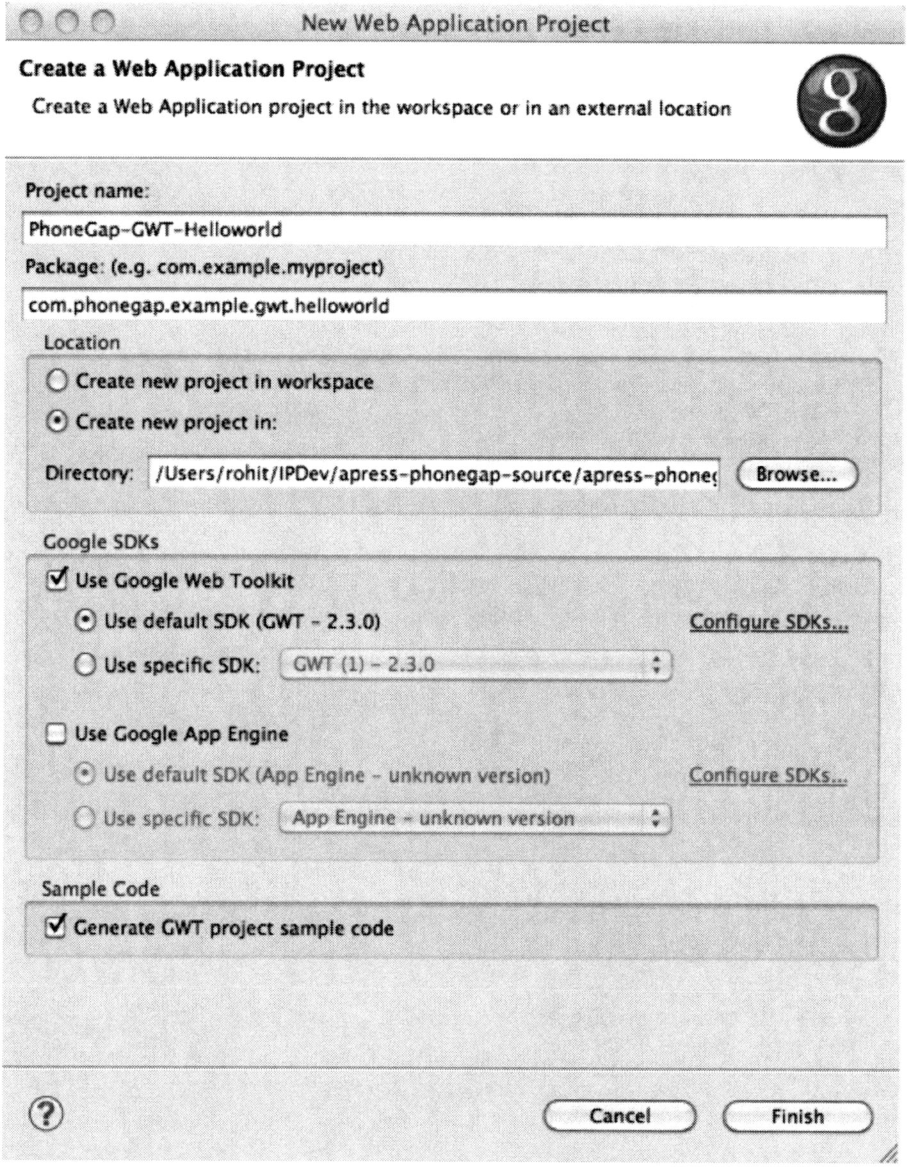

Figure 6–1. *Create GWT Project*

Create a folder named "lib" for the project. Download the GWT-PhoneGap library from http://code.google.com/p/gwt-phonegap/downloads/detail?name=gwt-phonegap-0.8.jar and copy it into the lib folder for your application. Add the gwt-phonegap-0.8.jar to the class path by right-clicking the jar and then clicking "build path" -> "Add to build path."

Now open PhoneGap_GWT_Helloworld.gwt.xml file. In that file, add the following entry:

```
<inherits name='de.kurka.phonegap.PhoneGap' />
<set-property name="user.agent" value="safari" />
```

Make a note that by adding the set-property of user.agent to safari, the GWT will only generate JavaScript for webkit-based browsers. Chrome browser will be used exclusively for testing in this scenario.

Your PhoneGap_GWT_Helloworld.gwt.xml file should now appear as follows:

```
<?xml version="1.0" encoding="UTF-8"?>
<module rename-to='phonegap_gwt_helloworld'>
  <!-- Inherit the core Web Toolkit stuff.                    -->
  <inherits name='com.google.gwt.user.User'/>
  <!-- Inherit the default GWT style sheet.  You can change   -->
  <!-- the theme of your GWT application by uncommenting       -->
  <!-- any one of the following lines.                         -->
  <inherits name='com.google.gwt.user.theme.clean.Clean'/>
  <!-- <inherits name='com.google.gwt.user.theme.standard.Standard'/> -->
  <!-- <inherits name='com.google.gwt.user.theme.chrome.Chrome'/> -->
  <!-- <inherits name='com.google.gwt.user.theme.dark.Dark'/>      -->

  <!-- Other module inherits                                   -->
  <inherits name='de.kurka.phonegap.PhoneGap' />
   <set-property name="user.agent" value="safari" />


  <!-- Specify the app entry point class.                      -->
  <entry-point
class='com.phonegap.example.gwt.helloworld.client.PhoneGap_GWT_Helloworld'/>
  <!-- Specify the paths for translatable code                 -->
  <source path='client'/>
  <source path='shared'/>

</module>
```

Now open PhoneGap_GWT_Helloworld.html, located in the war folder of your project, and make the following changes:

```
<!doctype html>
<!-- The DOCTYPE declaration above will set the     -->
<!-- browser's rendering engine into                 -->
<!-- "Standards Mode". Replacing this declaration    -->
<!-- with a "Quirks Mode" doctype may lead to some   -->
<!-- differences in layout.                          -->

<html>
  <head>
    <meta http-equiv="content-type" content="text/html; charset=UTF-8">

    <!--                                                        -->
    <!-- Consider inlining CSS to reduce the number of requested files -->
    <!--                                                        -->
    <link type="text/css" rel="stylesheet" href="PhoneGap_GWT_Helloworld.css">
```

```
    <!--                                  -->
    <!-- Any title is fine               -->
    <!--                                  -->
    <title>Gwt PhoneGap Demo</title>

    <!--                                  -->
    <!-- This script loads your compiled module.  -->
    <!-- If you add any GWT meta tags, they must   -->
    <!-- be added before this line.       -->
    <!--                                  -->
    <script type="text/javascript" language="javascript"
src="phonegap_gwt_helloworld/phonegap_gwt_helloworld.nocache.js"></script>
  </head>

  <!--                                  -->
  <!-- The body can have arbitrary html, or   -->
  <!-- you can leave the body empty if you want  -->
  <!-- to create a completely dynamic UI.    -->
  <!--                                  -->
  <body>

  </body>
</html>
```

If you plan to run this example on Android, you should make this addition after phonegap_gwt_helloworld/phonegap_gwt_helloworld.nocache.js tag:

```
<script type="text/javascript">
document.addEventListener("deviceready", (function(){ PhoneGap.available = true;}),
false);
</script>
```

Now open PhoneGap_GWT_Helloworld.java in the src folder and make the following changes:

```
package com.phonegap.example.gwt.helloworld.client;
import com.google.gwt.core.client.EntryPoint;
import com.google.gwt.user.client.ui.Label;
import com.google.gwt.user.client.ui.RootPanel;

/**
 * Entry point classes define <code>onModuleLoad()</code>.
 */
public class PhoneGap_GWT_Helloworld implements EntryPoint {

        /**
         * This is the entry point method.
         */
        public void onModuleLoad() {
                RootPanel.get().add(new Label("GWT PhoneGap Demo"));
        }
}
```

The GWT project created by default has an RPC component for client server communication, which is not needed for this application. Therefore, you can remove the following entries from the project:

- GreetingService.java and GreetingServiceAsync.java from the client package

- Shared and server packages

- Any servlets from web.xml

Now, run the GWT project (run as -> Web application) and you should see the screen shown in Figure 6–2. Please note that this example will be run in a browser to ensure that your GWT project is properly set up.

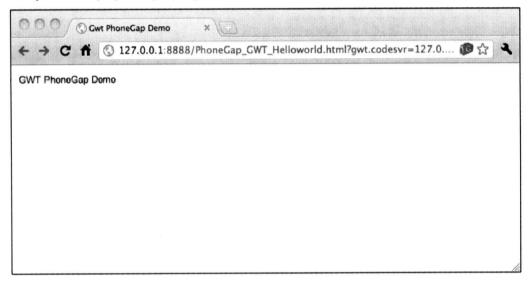

Figure 6–2. *Running GWT Project in Chrome Browser*

The next step is to actually make use of the PhoneGap API to compile the GWT project into a web application.

A benefit of the PhoneGap GWT library is that it mocks PhoneGap library when launched as a GWT web application. The library provides alternative functions based on the following instructions:

1. Use PhoneGap JavaScript if run on either Android or an iPhone.

2. Otherwise, use internal mock classes and give dummy values.

Start by using deferred binding to create an object of PhoneGap:

```
PhoneGap PhoneGap = (PhoneGap)GWT.create(PhoneGap.class);
```

The next step is to register the following callbacks within the PhoneGap framework:

- *Phonegapavailablehandler:* This callback will occur when everything goes fine and PhoneGap is initialized properly. In short, this is a success callback.

- *Phonegaptimeouthandler:* This callback occurs when PhoneGap is not initialized within the given time limit, possibly due to failure to initialize the PhoneGap framework. In short, this is a failure callback.

Lastly, you have to initialize the PhoneGap framework by calling `PhoneGap.initializePhoneGap()`. Calling this API will result in one of the above callbacks.

The main code will be written as the `PhoneGapAvailableHandler` Callback, as shown below. Using the PhoneGap variable is safe because PhoneGap has been properly initialized. In the following code, you get the handler to the device from PhoneGap and then print device info value in a grid (table of 2 columns and 5 rows):

```
Device device = phoneGap.getDevice();
Grid grid = new Grid(5, 2);
//Add a row mentioning Name Property of Device
grid.setWidget(0, 0, new Label("Name"));
grid.setWidget(0, 1, new Label(device.getName()));

//Add a row mentioning Platform Property of Device
grid.setWidget(1, 0, new Label("Platform"));
grid.setWidget(1, 1, new Label(device.getPlatform()));

//Add a row mentioning Version Property of Device
grid.setWidget(2, 0, new Label("Version"));
grid.setWidget(2, 1, new Label(device.getVersion()));

//Add a row mentioning Name Property of Device
grid.setWidget(3, 0, new Label("PhoneGapVersion"));
grid.setWidget(3, 1, new Label(device.getPhoneGapVersion()));

//Add a row mentioning Name Property of Device
grid.setWidget(4, 0, new Label("UUID"));
grid.setWidget(4, 1, new Label(device.getUuid()));

grid.setBorderWidth(1);
RootPanel.get().add(grid);
```

Here is the complete example:

```
package com.phonegap.example.gwt.helloworld.client;
import com.google.gwt.core.client.EntryPoint;
import com.google.gwt.core.client.GWT;
import com.google.gwt.user.client.Window;
import com.google.gwt.user.client.ui.Grid;
import com.google.gwt.user.client.ui.Label;
import com.google.gwt.user.client.ui.RootPanel;

import de.kurka.phonegap.client.PhoneGap;
import de.kurka.phonegap.client.PhoneGapAvailableEvent;
import de.kurka.phonegap.client.PhoneGapAvailableHandler;
import de.kurka.phonegap.client.PhoneGapTimeoutEvent;
```

```
import de.kurka.phonegap.client.PhoneGapTimeoutHandler;
import de.kurka.phonegap.client.device.Device;

/**
 * Entry point classes define <code>onModuleLoad()</code>.
 */
public class PhoneGap_GWT_Helloworld implements EntryPoint {

    /**
     * This is the entry point method.
     */
    public void onModuleLoad() {
        final PhoneGap phoneGap = GWT.create(PhoneGap.class);
        phoneGap.addHandler(new PhoneGapAvailableHandler() {

            public void onPhoneGapAvailable(PhoneGapAvailableEvent event) {
                Device device = phoneGap.getDevice();

                Grid grid = new Grid(5, 2);
                //Add a row mentioning Name Property of Device
                grid.setWidget(0, 0, new Label("Name"));
                grid.setWidget(0, 1, new Label(device.getName()));
                //Add a row mentioning Platform Property of Device
                grid.setWidget(1, 0, new Label("Platform"));
                grid.setWidget(1, 1, new Label(device.getPlatform()));
                //Add a row mentioning Version Property of Device
                grid.setWidget(2, 0, new Label("Version"));
                grid.setWidget(2, 1, new Label(device.getVersion()));
                //Add a row mentioning Name Property of Device
                grid.setWidget(3, 0, new Label("PhoneGapVersion"));
                grid.setWidget(3, 1, new Label(device.getPhoneGapVersion()));
                //Add a row mentioning Name Property of Device
                grid.setWidget(4, 0, new Label("UUID"));
                grid.setWidget(4, 1, new Label(device.getUuid()));
                grid.setBorderWidth(1);
                RootPanel.get().add(grid);

            }
        });

        phoneGap.addHandler(new PhoneGapTimeoutHandler() {
            public void onPhoneGapTimeout(PhoneGapTimeoutEvent event) {
                Window.alert("can not load phonegap");
            }
        });

        phoneGap.initializePhoneGap();
    }
}
```

You can run this example from Eclipse using "run as -> web application" and looking up the code in the browser at
http://127.0.0.1:8888/PhoneGap_GWT_Helloworld.html?gwt.codesvr=127.0.0.1:9997.

You will see the table shown in Figure 6–3. As mentioned above, mock values are shown by the GWT PhoneGap when running the code in a browser and not on Android or an iPhone.

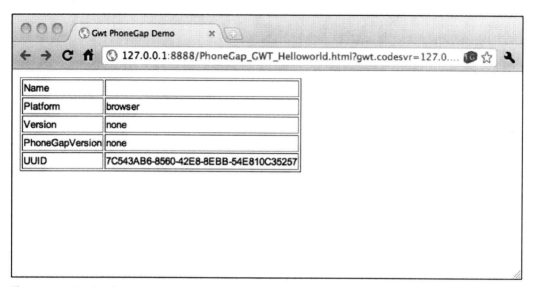

Figure 6–3. *Running GWT PhoneGap Project in Chrome Browser*

The last step is to compile this project into a web application. Right-click the project, choose the Google option, and click the menu option named "GWT compile." You will be presented with the dialog box shown in Figure 6–4. Click compile.

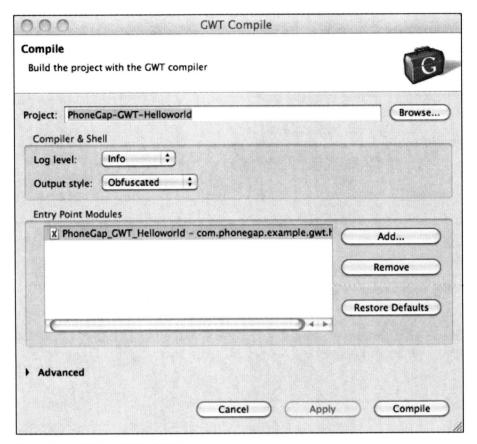

Figure 6–4. *GWT Compilation Screen*

When the compilation is done, refresh your project in Eclipse, and you should see the directory structure shown in Figure 6–5. Inside the war folder you should see a phonegap_gwt_helloworld folder containing many HTML and JavaScript files, as shown below.

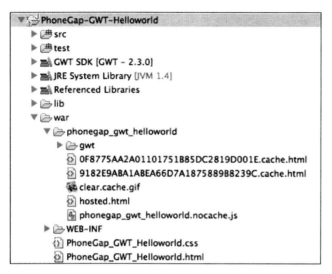

Figure 6–5. *GWT War Directory Structure after GWT Compilation*

Build a PhoneGap Android Application

The final step in building the PhoneGap GWT application is to create an Android PhoneGap project, as shown in Figure 6–6 and Figure 6–7, and then copy the GWT-generated web application into the assets/www folder.

The first step is to create an Android project.

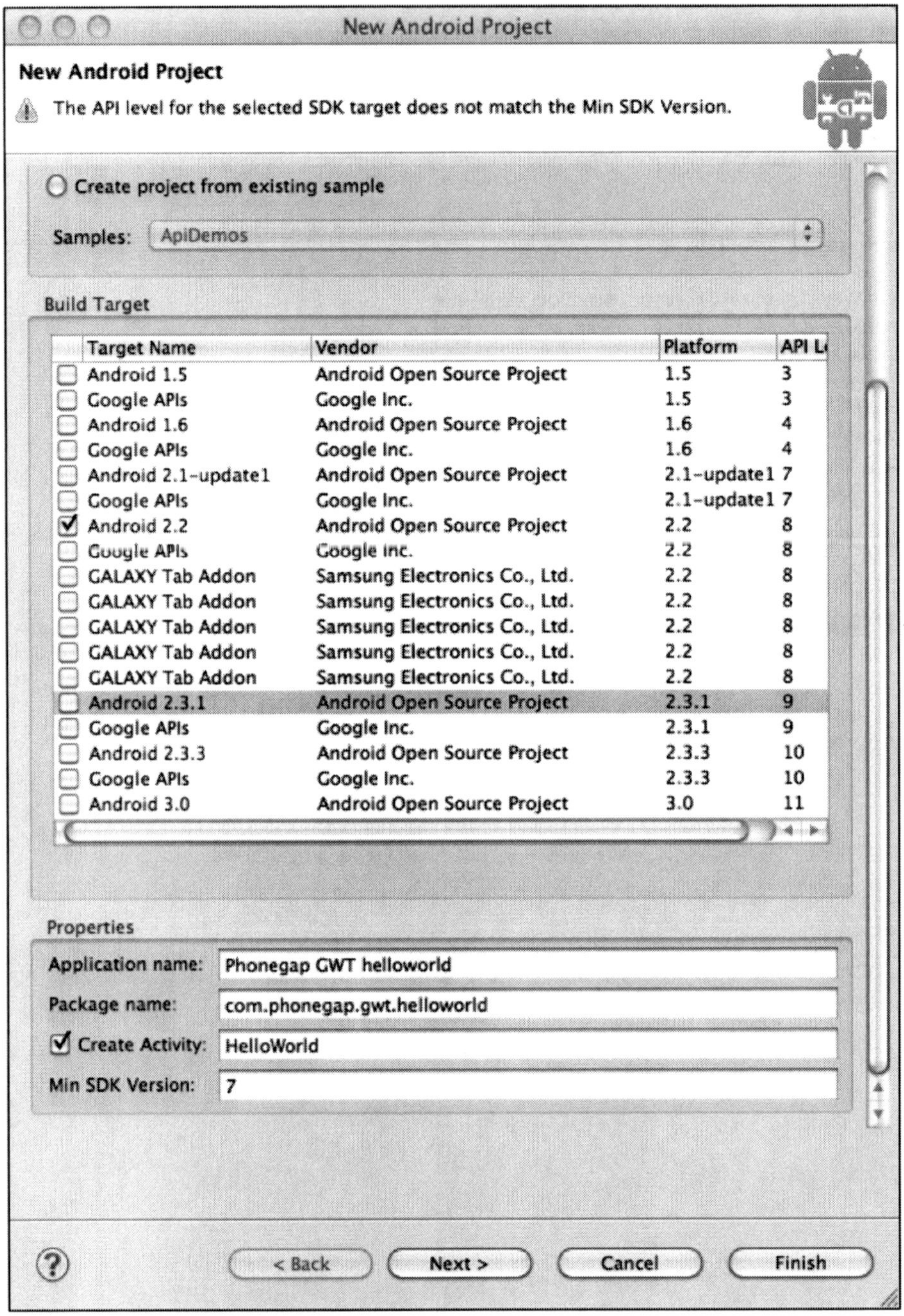

Figure 6–6. *Android Create Project Screen*

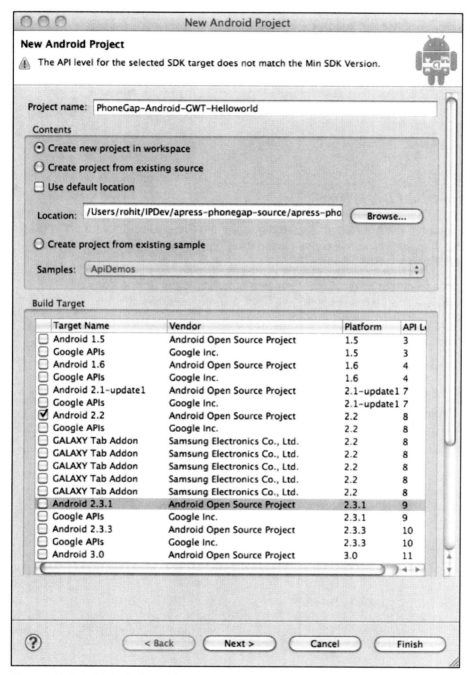

Figure 6–7. *Android Create Project Screen*

Now, inject the PhoneGap 0.9.4 library into the Android project.

Download the PhoneGap 0.9.4 library from
`http://phonegap.googlecode.com/files/phonegap-0.9.4.zip`. Undo the zip file, and you
will see the folder structure shown in Figure 6–8.

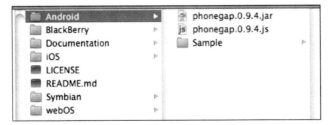

Figure 6–8. *PhoneGap 0.9.4 Directory Structure*

Create a lib folder in the Android project folder, copy the PhoneGap.0.9.4.jar file in the
lib folder, and then add it to the Eclipse classpath. (Right-click the jar file, go to "Build
Path," and click "Add to Build Path.")

The next step is to create a www folder inside the assets folder and to copy the
PhoneGap.0.9.4.js file into the www folder. Then you will need to copy the following files
from the GWT project into the same folder:

- PhoneGap_GWT_Helloworld.html

- PhoneGap_Gwt_Helloworld.css

- phonegap_gwt_helloworld folder

Your folder structure should now look like the example in Figure 6–9. The files in the
folder named "gwt" are generated when the project is compiled.

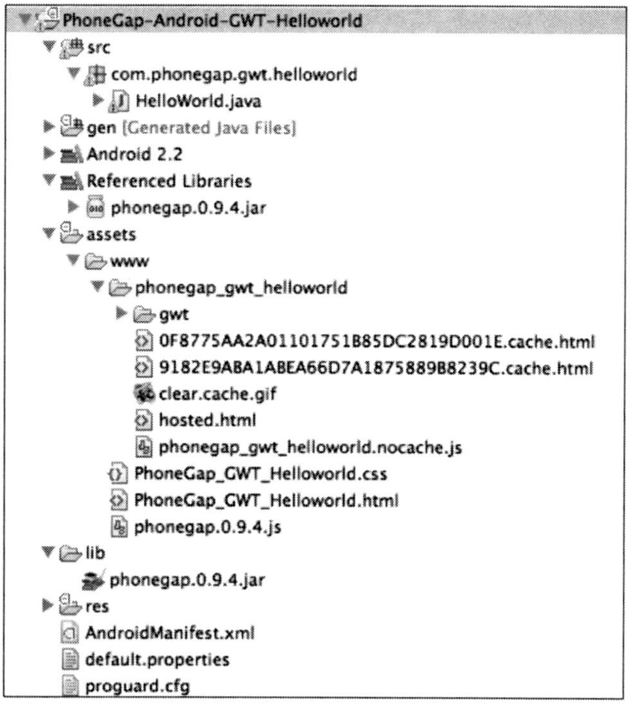

```
▼ ⚏ PhoneGap-Android-GWT-Helloworld
    ▼ ⚏ src
        ▼ ⚏ com.phonegap.gwt.helloworld
            ▶ ⚏ HelloWorld.java
    ▶ ⚏ gen [Generated Java Files]
    ▶ ⚏ Android 2.2
    ▼ ⚏ Referenced Libraries
        ▶ ⚏ phonegap.0.9.4.jar
    ▼ ⚏ assets
        ▼ ⚏ www
            ▼ ⚏ phonegap_gwt_helloworld
                ▶ ⚏ gwt
                    ⚏ 0F8775AA2A01101751B85DC2819D001E.cache.html
                    ⚏ 9182E9ABA1ABEA66D7A1875889B8239C.cache.html
                    ⚏ clear.cache.gif
                    ⚏ hosted.html
                    ⚏ phonegap_gwt_helloworld.nocache.js
                ⚏ PhoneGap_GWT_Helloworld.css
                ⚏ PhoneGap_GWT_Helloworld.html
                ⚏ phonegap.0.9.4.js
    ▼ ⚏ lib
        ⚏ phonegap.0.9.4.jar
    ▶ ⚏ res
        ⚏ AndroidManifest.xml
        ⚏ default.properties
        ⚏ proguard.cfg
```

Figure 6–9. *Directory Structure of GWT PhoneGap Project using Android*

Now you need to modify the following files:

- HelloWorld.java file

- PhoneGap_GWT_Helloworld.html

Also, you will need o check that HelloWorld.java file resembles the following:

```java
package com.phonegap.gwt.helloworld;
import android.os.Bundle;
import com.phonegap.DroidGap;
public class HelloWorld extends DroidGap {
    /** Called when the activity is first created. */
    @Override
    public void onCreate(Bundle savedInstanceState) {
        super.onCreate(savedInstanceState);
        super.loadUrl("file:///android_asset/www/PhoneGap_GWT_Helloworld.html");
    }
}
```

In the PhoneGap_GWT_Helloworld.html folder, you will need to make some code changes.

First, add the PhoneGap JavaScript library version 0.9.4 for Android, as follows:

```
<script type="text/javascript" src="phonegap.0.9.4.js"></script>
```

Next, listen to the deviceready event in order to determine whether the PhoneGap library is ready for use:

```
<script type="text/javascript">
    document.addEventListener(
        "deviceready",
        (function() {
            PhoneGap.available = true;
        }),
        false);
</script>
```

Explicitly set the PhoneGap.available variable to true here. This is a required step for the Android platform.

Here is the complete source code for the PhoneGap_gWT_Helloworld.html:

```
<!doctype html>
<!-- The DOCTYPE declaration above will set the -->
<!-- browser's rendering engine into -->
<!-- "Standards Mode". Replacing this declaration -->
<!-- with a "Quirks Mode" doctype may lead to some -->
<!-- differences in layout. -->
<html>

    <head>
        <meta http-equiv="content-type" content="text/html; charset=UTF-8">
        <!-- -->
        <!-- Consider inlining CSS to reduce the number of requested files -->
        <!-- -->
        <link type="text/css" rel="stylesheet" href="PhoneGap_GWT_Helloworld.css">
        <!-- -->
        <!-- Any title is fine -->
        <!-- -->
        <title>
            Gwt PhoneGap Demo
        </title>
        <!-- -->
        <!-- This script loads your compiled module. -->
        <!-- If you add any GWT meta tags, they must -->
        <!-- be added before this line. -->
        <!-- -->
        <script type="text/javascript" language="javascript" src="phonegap.0.9.4.js">
        </script>
        <script type="text/javascript" language="javascript"
src="phonegap_gwt_helloworld/phonegap_gwt_helloworld.nocache.js">
        </script>
        <script type="text/javascript">
                    document.addEventListener("deviceready", (function() {
                PhoneGap.available = true;
            }), false);
        </script>
    </head>
    <!-- -->
```

```
<!-- The body can have arbitrary html, or -->
<!-- you can leave the body empty if you want -->
<!-- to create a completely dynamic UI. -->
<!-- -->

<body>
</body>
```

```
</html>
```

After you run this code, the screen on the emulator should appear as the example in
Figure 6–10.

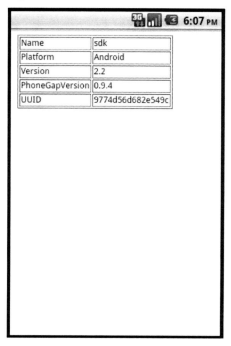

Figure 6–10. *GWT based PhoneGap Application showing Device Info*

As with the above code, you can write code to access other PhoneGap APIs, and you
can also write a GWT application that accesses native phone features via PhoneGap.

GWT PhoneGap Reference

Links to the locations of the documentation and the source code used in this GWT
PhoneGap project are listed below. Daniel Kurka is the author of this library.

Home page – `http://code.google.com/p/gwt-phonegap/`

Getting started – `http://code.google.com/p/gwt-phonegap/wiki/GettingStarted`

Download jar – `http://gwt-phonegap.googlecode.com/files/gwt-phonegap-0.8.jar`

Download Javadocs – `http://gwt-phonegap.googlecode.com/files/gwt-phonegap-0.8-javadoc.jar`

Source code – `http://code.google.com/p/gwt-phonegap/source/browse/`

Current features – `http://code.google.com/p/gwt-phonegap/wiki/Features`

Chapter 7

PhoneGap Emulator and Remote Debugging

Introduction

The biggest pain I experienced when I worked on building the PhoneGap application is the following cycle:

1. Develop an eclipse/xcode or irrespective IDE

2. Compile and put the binary executable on device/emulator

3. Test PhoneGap application on device/emulator

4. Tweak code and repeat from step 1

Clearly this cycle is very time consuming and frustrating. If you are an experienced JavaScript developer, this would be a nightmare for you.

JavaScript developers are used to handy tools on the following modern browsers:

1. Firefox

2. Safari

3. Chrome

4. Internet Explorer

(Note that for iPhone and Android development, Internet Explorer is not useful. We recommend the use of Firefox, Safari, or Chrome for iPhone and Android development.)

Few of these tools are extensions for developers. Firefox has its own firebug, which is the first of the series, of javascript/html debugging tools and it allows you to debug not only the page elements (DOM structure), but the scripts, stylesheets, and network as well. It also allows you to change these things on fly and test them out immediately.

Chrome and Safari come with built-in developer tools. Internet Explorer is no exception and has its own extension for doing similar things.

Clearly we need something similar while developing PhoneGap. Let's list our two requirements here. We need the following:

1. The ability to create and test applications outside PhoneGap in a browser world using a PhoneGap emulator

2. The ability to debug PhoneGap applications once we have deployed them on some emulator or device

In this chapter we will discuss using a PhoneGap emulator and remote debugging tool.

PhoneGap Emulator for Chrome – Using Ripple

Ripple is a multi-platform mobile platform emulator from a company named tinyHippos. Recently this company was acquired by Research In Motion (RIM). The main reason why Ripple came into existence was to reduce the challenges being faced by today's mobile web developers due to the immense fragmentation in the mobile OS world.

Ripple is a Chrome extension and provides simulation for the following:

1. PhoneGap

2. Webworks (from blackberry)

3. WebWorks-Tablet-OS (from Blackberry)

4. Mobile web

5. WAC

6. Opera

7. Vodafone

For the purposes of this book, we will only focus on PhoneGap emulation from Ripple.

Installing Ripple

The only prerequisite for Ripple is that you need the Chrome browser. Any version that supports extensions will do, so you don't need to worry about which version of Chrome to install. However, we do recommend using the latest version.

Open Chrome and visit the web site – `http://ripple.tinyhippos.com/`

You will see the page shown in Figure 7–1; all you need to do is click on the "Get Ripple" button.

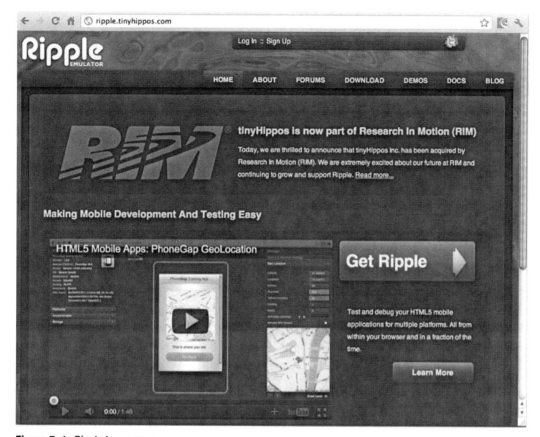

Figure 7–1. *Ripple home page*

That will take you to the page shown in Figure 7–2. Now click on "Install."

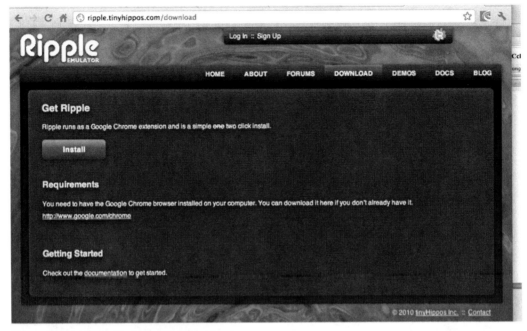

Figure 7–2. *Installing Ripple as a Chrome extension*

As we mentioned earlier, Ripple is a Chrome extension. Clicking "Install" takes you to the Chrome web store. When you click "Add to Chrome," the extension/plugin is actually downloaded from the Chrome web store and automatically installed in your Chrome browser (see Figure 7–3).

Figure 7–3. *Installing Ripple from Chrome web store*

Once the plugin is installed you will see a screen like the one shown in Figure 7–3.

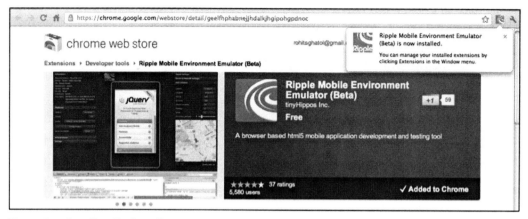

Figure 7–4. *Installing Ripple on Chrome*

In order to verify that the plugin has been properly installed, open a site like www.google.com in Chrome and right click on the page. If the plugin has been properly installed, you should see an option for "Emulator" enable/disable. This is depicted in Figure 7–5.

Figure 7–5. *Right click option on Chrome to open the Ripple emulator*

Go ahead and click on "Emulator" and select "Enable." If you see the screen in Figure 7–6, your Ripple plugin is working perfectly. To get out of that screen, simply right click and select Emulator - >Disable.

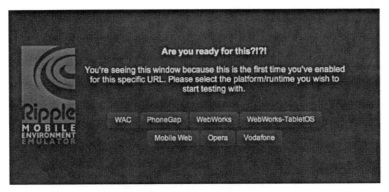

Figure 7–6. *First time launch of Ripple*

Using Chrome Effectively for PhoneGap

Before we go ahead and use Ripple on Chrome. Let's understand how to use Chrome effectively for PhoneGap. PhoneGap has similarities with mobile web applications, but it differs in many areas as well. Let's list out the two major differences.

1. PhoneGap applications are HTML/JavaScript-based applications, but they are never hosted. They are bundled as a part of native mobile applications shown in an embedded browser inside the application. This is similar to testing your HTML/JavaScript application from your disk on Chrome. Chrome needs to be tweaked to support running html/javascript applications from disk.

2. PhoneGap does not have any domain name associated with them. That is why they do not follow a single origin policy. If you want to simulate the not adhering to single origin policy for your PhoneGap application, which is hosted from file system or local web server, you need to tweak Chrome to support turning off single origin policy.

In order to test the PhoneGap application from the local file system and not make it follow the single origin policy, you need to start Chrome with the command line arguments discussed in the next few paragraphs.

Windows

In Windows, create a .cmd file named chrome.cmd and copy the following script into that file. Now use chrome.cmd to launch Chrome.

```
chrome.exe --disable-web-security --allow-file-access-from-files
```

Mac and Linux

On a Mac, create a script named chrome.sh and copy the following script into that. Now use chrome.sh to launch Chrome.

```
open "/Applications/Google Chrome.app" --args --disable-web-security --allow-file-
access-from-files
```

Change the permissions on the chrome.sh script to make it executable (needed to run the script).

```
$>chmod +x chrome.sh
Run the script from terminal as follows:
$>./chrome.sh
```

Using Ripple

Now we will see how to use Ripple and what changes are required to run your PhoneGap app in Ripple.

Following are the prerequisites for using a PhoneGap app in Ripple:

1. The app needs to be a pure PhoneGap app with no plugins.

2. You need to remove any reference to the PhoneGap JavaScript from all HTML files.

Tune Your App for Ripple

Let's take a code example from Chapter 2.

We will work on the example of a compass app. Take the compass image from this URL – http://beginingphonegap.googlecode.com/files/compass.png. This image is shown in Figure 7–7.

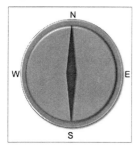

Figure 7–7. *Compass image to be used in PhoneGap application*

Now, let's modify the index.html file to look as it does below. Make a note that we have removed any reference to phonegap.js. This is currently a prerequisite for using Ripple. Ripple is working with PhoneGap to get rid of this change. Hopefully, in upcoming releases, we will see this requirement disappear.

```
<!DOCTYPE HTML>
<html>

    <head>
        <title>
            PhoneGap
        </title>
```

```html
<script type="text/javascript">
        /** Called when phonegap javascript is loaded */

function onDeviceReady() {
    var button = document.getElementById("capture");
    var compassOptions = {
        frequency: 1000
    };
    navigator.compass.watchHeading(onSuccess, onError, compassOptions);
};

function onSuccess(heading) {
    var image = document.getElementById('compass');
    var headingDiv = document.getElementById('compassHeading');
    headingDiv.innerHTML = heading;
    var reverseHeading = 360 - heading;
    image.style.webkitTransform = "rotate(" + reverseHeading + "deg)";
}

function onError(error) {
    alert('code: ' + error.code + '\n' + 'message: ' + error.message + '\n');
}

  /** Called when browser load this page*/

function init() {
    document.addEventListener("deviceready", onDeviceReady, false);
}
    </script>
</head>

<body onLoad="init()">
    <h1>
        Compass
    </h1>
    <table>
        <tr>
            <td>
                Compass Heading
            </td>
            <td>
                <div id="compassHeading">
                    ....
                </div>
            </td>
            <td>
                Degrees
            </td>
        </tr>
    </table>
    <img id="compass" src="compass.png" style="width:400px;height:400px;margin-
left:auto;margin-right:auto;auto;display:block">
    </img>
</body>

</html>
```

Your app directory would look like Figure 7–8. It contains all HTML, JavaScript, and CSS files for your application.

Figure 7–8. *Application Directory for the Compass App*

Start Chrome with Special Flags

The next step is to start Chrome with special flags (as we started earlier in this chapter).

Start Chrome with --disable-web-security --allow-file-access-from-files flags.

Once Chrome starts, go to Window->Extensions and locate the "Ripple Mobile Environment Emulator" extension and enable the check box that says "Allow access to file URLs" (see Figure 7–9).

Figure 7–9. *Allow access to file URLs for Ripple extension*

Load App in Chrome

Now load the compass application in Chrome. On the right-hand side in the top corner you will see the Ripple icon. Click on that item to enable Ripple for this app. This is shown in Figure 7–10.

Make note because we started Chrome with above mentioned flags, that is why Chrome is able to load HTML files from local filesystem. Also, if the PhoneGap application loads data using Ajax, it would also work in Chrome as we have disabled the single origin policy.

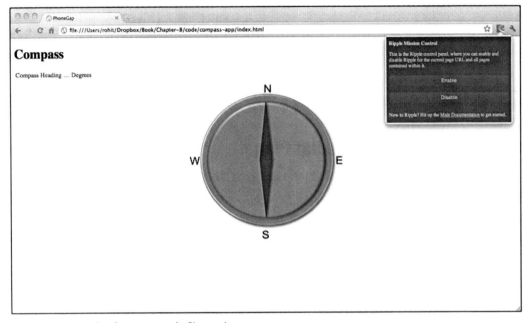

Figure 7–10. *Loading Compass app in Chrome browser*

Enable Ripple

You will see Figure 7–11 the first time you enable Ripple. We need to choose the PhoneGap option.

Figure 7–11. *Enabling Ripple for Compass app*

Play with Ripple Settings

Now that we have enabled Ripple, we will see the webpage has changed. The application, which used to take up the entire screen, now looks different. This is because now Ripple is the main application. Ripple loads our own application in an iframe and injects itself to simulate PhoneGap Environment. On the main page, Ripple has given a number of controls to change the state and properties of devices that PhoneGap simulates (see Figure 7–12).

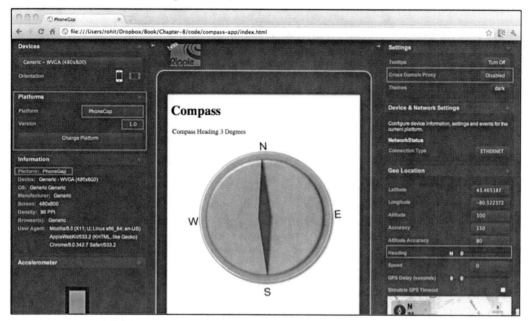

Figure 7–12. *Compass app loaded in Chrome browser with Ripple enabled*

Test Application with Ripple

The application under test is a compass application. To test this application we will use the geo and compass control marked in red in the bottom-right corner (see Figure 7–12). If we change the heading, this means we are simulating the user moving the compass bearing of his/her device.

As you can see in Figure 7–13, when we changed the heading we can see the compass image rotate around center.

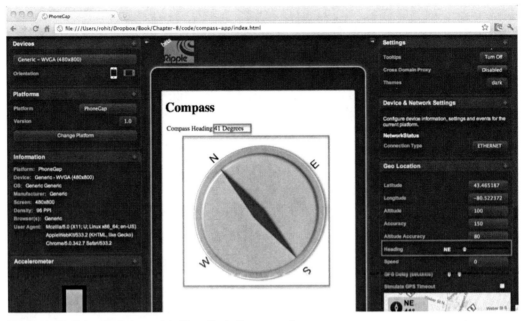

Figure 7–13. *Using Ripple to emulate PhoneGap's Compass api*

Simulating PhoneGap is just one angle of Ripple. Ripple allows developers to use their regular browser tools to debug and change the DOM and CSS of their application. In Figure 7–14, we have right clicked on the image and clicked on "Inspect Element." We can then check the css style of inspected DOM element, in our case the html img.

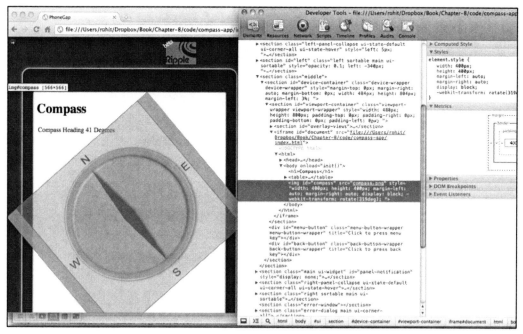

Figure 7–14. *Using Chrome's developer tool to see HTML DOM changes*

Remote Debugging – http://debug.phonegap.com

While using Ripple to simulate, test, and debug PhoneGap applications works and is very helpful, nothing compares to debugging on an actual emulator or device.

The problem with debugging on an actual emulator or device is that the webkit webview used to show the PhoneGap application is pretty much isolated and cannot be accessed from the outside. Compare this to Chrome, Firefox or Safari, in these browsers user can inspect an html element. But the Applications (HTML/Javascript) which run inside a WebView of say PhoneGap Applications cannot be inspected.

Here is where remote debugging comes into picture. See Figure 7–15 to understand this concept better. The basic idea is to inject a debug JavaScript into our PhoneGap application. This opens a channel with the debug.phonegap.com server. The developer then opens debug.phonegap.com in a browser and inspects the PhoneGap application running on the device/emulator.

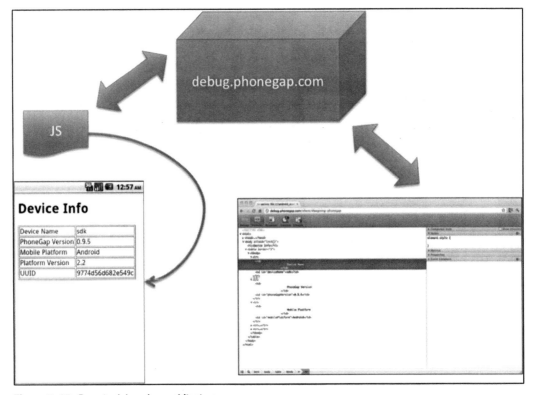

Figure 7–15. *Remote debugging architecture*

Setting up Remote Debugging

The first step to do a remote debug is to open http://debug.phonegap.com in a browser.
Here you can provide a guide of your own (like we did) or use the one the server
randomly assigns.

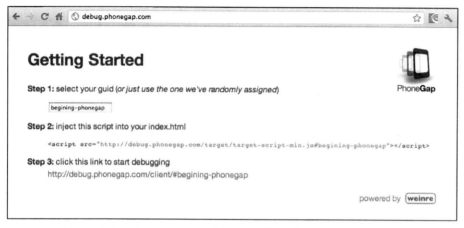

Figure 7–16. *Using debug.phonegap.com to debug your PhoneGap application*

Injecting Remote Debugging in the PhoneGap App

The next step is to copy the JavaScript snippet from `http://debug.phonegap.com` and inject it into your PhoneGap application. This is shown as underlined below.

```
<!DOCTYPE HTML>
<html>
    <head>
        <title>PhoneGap</title>
        <script type="text/javascript" src="phonegap.1.1.0.js">
        </script>
        <script type="text/javascript">

            /** Called when phonegap javascript is loaded */
            function onDeviceReady(){
                document.getElementById("deviceName").innerHTML = device.name;
                document.getElementById("phoneGapVersion").innerHTML = device.phonegap;
                document.getElementById("mobilePlatform").innerHTML = device.platform;
                document.getElementById("platformVersion").innerHTML = device.version;
                document.getElementById("uuid").innerHTML = device.uuid;
            }

            /** Called when browser load this page*/
            function init(){
                document.addEventListener("deviceready", onDeviceReady, false);
            }
        </script>
        <script src="http://debug.phonegap.com/target/target-script-min.js#begining-
phonegap">
        </script>
    </head>
    <body onLoad="init()">
        <h1>Device Info</h1>
        <table border="1">
            <tr>
                <td>
                    Device Name
```

```
            </td>
            <td id="deviceName">
            </td>
        </tr>
        <tr>
            <td>
                PhoneGap Version
            </td>
            <td id="phoneGapVersion">
            </td>
        </tr>
        <tr>
            <td>
                Mobile Platform
            </td>
            <td id="mobilePlatform">
            </td>
        </tr>
        <tr>
            <td>
                Platform Version
            </td>
            <td id="platformVersion">
            </td>
        </tr>
        <tr>
            <td>
                UUID
            </td>
            <td id="uuid">
            </td>
        </tr>
    </table>
  </body>
</html>
```

Debugging and Modifying the DOM Element

The next step is to launch the PhoneGap application on an emulator or device. When this application starts, the JavaScript running inside it will communicate with the debug.phonegap.com server. Then we will be ready for remote debugging (see Figure 7–17).

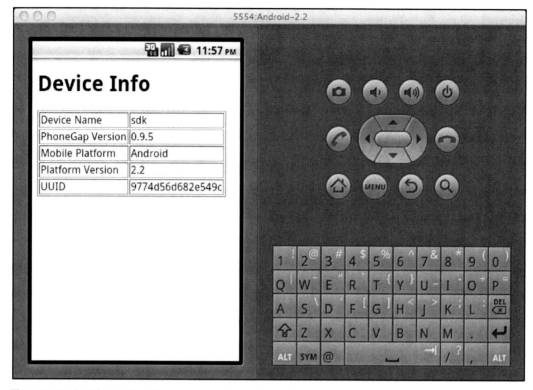

Figure 7–17. *Loading device info application in Android emulator*

The last step is to open `http://debug.phonegap.com/client/#begining-phonegap` in a browser (see Figure 7–18). Remember, we got this URL from the `http://debug.phonegap.com` website.

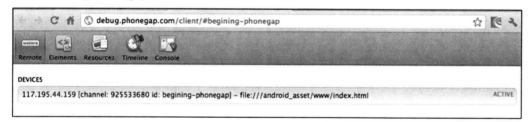

Figure 7–18. *Log message showing an Android application connected to http://debug.phonegap.com*

Now move to the element tab (see Figure 7–19). Here you will be able to see the DOM element of the page in the PhoneGap application.

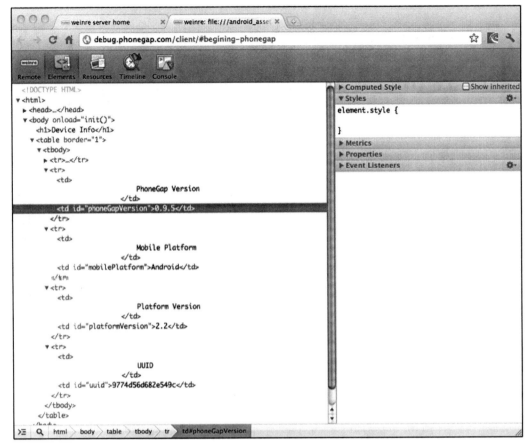

Figure 7–19. *Inspecting DOM on http://debug.phonegap.com*

The fun doesn't stop here. Now you can change the DOM element (see Figure 7–20). To do so, double click on any DOM element (in our case a TD) and type in the style part by hand. In our case, we will add a style "style='background:red'" to td containing 0.9.5 text. Now we will switch to the PhoneGap application running in the emulator to see the changes coming into effect.

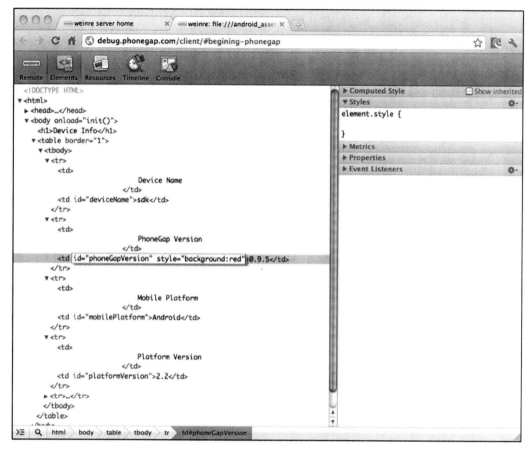

Figure 7–20. *Changing a DOM property on http://debug.phonegap.com*

Now the background color of the TD element containing "0.9.5" changes to red (see
Figure 7–21). This kind of debugging helps us debug applications on a device/emulator
in real time.

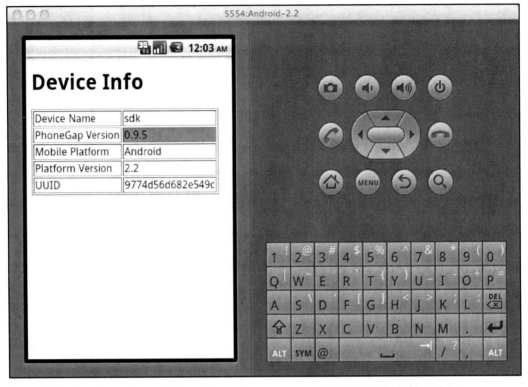

Figure 7–21. *The change done on* `http://debug.phonegap.com` *reflected on an Android emulator*

Issues with debug.phonegap.com

So far we've seen that the use of `http://debug.phonegap.com` is pretty useful in inspecting what is running inside a real device or an emulator. However, we won't want to use it for following reasons:

1. During development we don't want to use an outside server.

2. We want to save bandwidth and increase the speed of debugging.

Note: A Weinre (**we**b **in**spector **re**mote) server powers debug.phonegap.com. PhoneGap folks also developed Weinre, and they have documented it pretty well.

Installing Local debug.phonegap.com

While installations and deployment of Weinre is out of scope of this book, I will leave a few instructions in the form of links on how to locally deploy Weinre.

The documentation for Weinre is pretty good, and if you follow it you should have no problem locally installing Weinre and using it.

Visit http://phonegap.github.com/weinre/Installing.html for installation documentation (see Figure 7–22).

Figure 7–22. *Instructions on how to install Weinre*

Visit http://phonegap.github.com/weinre/Running.html for instructions on how to run it (see Figure 7–23).

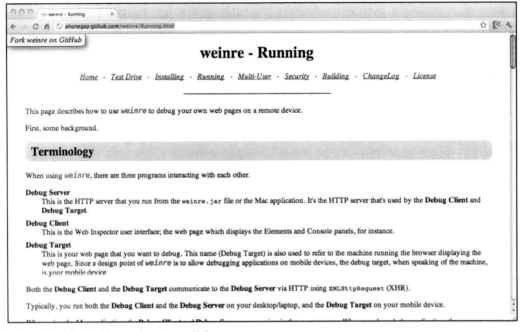

Figure 7–23. *Instructions on how to run Weinre*

Conclusion

When developing PhoneGap applications, using an iPhone/Android emulator to test code is very time consuming and frustrating. In order to save time and energy and ease out development, use the Ripple PhoneGap Emulator. Make the most of Chrome's developer tools and speed up your development. For remote debugging, use either `http://debug.phonegap.com` or a locally installed Weinre server. This helps you know what is actually happening to the HTML DOM when the application is running on the iPhone/Android emulator or on the actual device.

Chapter 8

Using PhoneGap Plug-Ins

PhoneGap comes with a set of JavaScript APIs that are used to access native phone features, such as camera, storage, contacts, geolocation, etc., to build cross mobile applications. If you want to do something that is not available in a PhoneGap API, you can leverage the PhoneGap plug-in.

In any technology, it's common practice to reuse a feature that is already available and tested. There are significant third party plug-ins available for the PhoneGap. For example, the authentication mechanism to access Facebook, access a third party service for Mobile Push notification, etc.

What Is PhoneGap Plug-In?

PhoneGap plug-in is an extension of the PhoneGap feature. It accesses a piece of functionality on the phone. Plug-in functionality may only be able to access native features of the phone or it may provide the functionality to access cloud services.

Any PhoneGap Plug-in consists of at least two files:

- JavaScript file
- Native language file

Plug-In's JavaScript file is the interface between the PhoneGap's application and the PhoneGap's plug-in. Plug-In's functionalities are accessed by the JavaScript file using JavaScript functions.

A native language file is used by the PhoneGap framework to interact with the phone to access native features. As a plug-in user, we need to place the native code into our project structure. In the next section, we will examine in detail how to set up the project using this plug-in.

Facebook Authentication and Fetching Friends

Let's leverage the PhoneGap plug-in to build a small application to login to Facebook and fetch friends' info from Facebook.

Our PhoneGap application will use the Facebook native app to perform a single sign on (SSO) for the user through the Facebook-PhoneGap plug-in.

Setting Environment for Android

First, we will need to set up the PhoneGap project for Android. Please refer to Chapter 2 to setup your project for Android. An Android project configuration is shown in Figure 8–1.

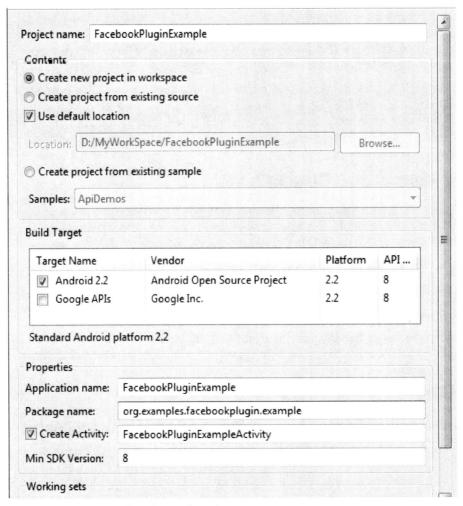

Figure 8–1. *Eclipse Android project configuration*

Download the Facebook-connect plug-in from
`https://github.com/davejohnson/phonegap-plugin-facebook-connect/downloads`.

The Facebook-connect plug-in is a zip file. Unzip it in your favorite folder. The folder
structure should be similar to Figure 8–2.

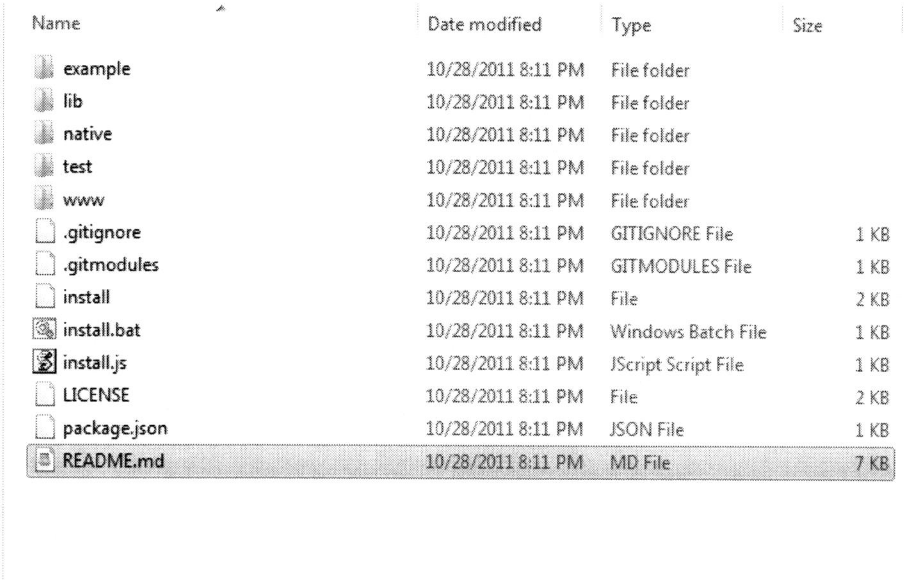

Name	Date modified	Type	Size
example	10/28/2011 8:11 PM	File folder	
lib	10/28/2011 8:11 PM	File folder	
native	10/28/2011 8:11 PM	File folder	
test	10/28/2011 8:11 PM	File folder	
www	10/28/2011 8:11 PM	File folder	
.gitignore	10/28/2011 8:11 PM	GITIGNORE File	1 KB
.gitmodules	10/28/2011 8:11 PM	GITMODULES File	1 KB
install	10/28/2011 8:11 PM	File	2 KB
install.bat	10/28/2011 8:11 PM	Windows Batch File	1 KB
install.js	10/28/2011 8:11 PM	JScript Script File	1 KB
LICENSE	10/28/2011 8:11 PM	File	2 KB
package.json	10/28/2011 8:11 PM	JSON File	1 KB
README.md	10/28/2011 8:11 PM	MD File	7 KB

Figure 8–2. *Facebook-connect plug-in folder structure*

Next you will need to perform the following installation steps:

1. Register Facebook Plug-in

   Add the following XML element as a child of the "plugins" element in the
   plugins.xml file as shown in Figure 8–3. You may need to create an xml folder in
   the res folder using the following:

```
<plugin name="com.phonegap.facebook.Connect" value="com.phonegap.facebook.ConnectPlugin" />
```

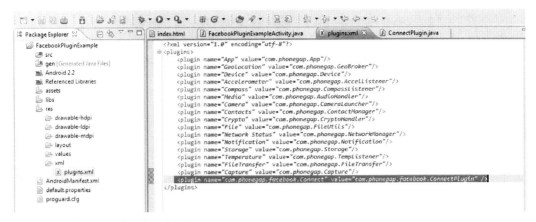

Figure 8–3. *Facebook plug-in registration*

2. Include the native part of the plug-in into the project. Copy the libs and src folder from a Facebook-connect-plug-in folder as shown in Figure 8–4 and paste it into the root of our PhoneGap application, i.e., "FaceBookPluginExample".

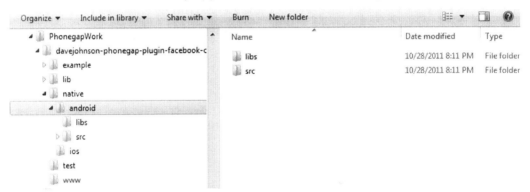

Figure 8–4. *Facebook-connect-plug-in native folders for android*

3. Include the JavaScript part of the plug-in in the project

There are two JavaScript files that we need to include in our project from Facebook-connect plug-in.

- /www/pg-plugin-fb-connect.js

   pg-plugin-fb-connect.js file is available under the www folder of the Facebook plug-in. Copy and paste it into the assets/www folder of our project.

- /lib/facebook_js_sdk.js

   facebook_js_sdk.js file is available under the lib folder of the Facebook plug-in. Copy and paste it into the assets/www folder of our project.

Once you are done with these three steps, you will see the FaceBookPluginExample project structure as shown in Figure 8–5.

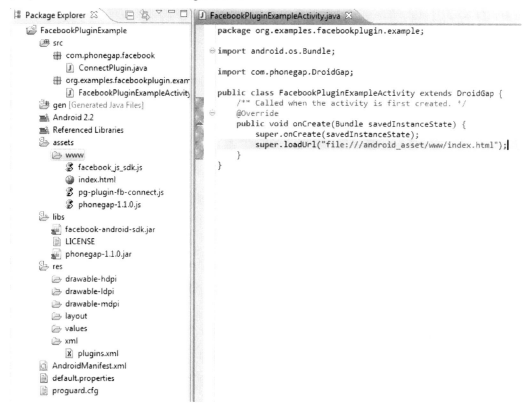

Figure 8–5. *FaceBookPluginExample project structure*

Initializing Facebook-Connect Plug-In

The first step is to make sure that index.html has a Facebook-Connect library, PhoneGap library, and is CSS linked. Note that we are including the following JavaScript files.

1. PhoneGap JavaScript

2. Facebook Plug-In JavaScript

3. Facebook SDK JavaScript

```html
<html>
<head></head>
<body>
<div id="friends"></div>
<!--phonegap -->
<script src="phonegap-1.1.0.js"></script>
<!--phonegapfacebook plugin -->
<script src="pg-plugin-fb-connect.js"></script>
<!--facebookjssdk -->
```

```
<script src="facebook_js_sdk.js"></script>
</body>
</html>
```

Now we will define the JavaScript functions in the index.html page to login into Facebook and fetch the friend list. The following is the code snip for the login function:

```
function login() {
    FB.login(function(response) {…},
                    { perms: "email" }
            );
}
```

login() function calls the Facebook SDK's login function FB.login(). Facebook's FB.login() has two parameters. The first is the callback JavaScript function and the second is the JSON object, which is used to specify permissions. We are passing 'function(response){…}' and '{ perms: "email" }' into FB.login(). Facebook's FB.login() prompts the user to login. Upon successful login, it calls the callback JavaScript function. The callback function gets the 'response' object to identify the login status. 'perms' is used to specify the user permission. You can find more detail about the Facebook login API and user permissions from the Facebook developer site at http://developers.facebook.com/docs/reference/api/permissions.

Next, we will see the code snippet that is used to get the friend list. To do so, we will create a JavaScript function getFriendList().

```
function getFriendList(){
    FB.api('/me/friends', function(response) {
            if (response.error) {
    alert(JSON.stringify(response.error));
            } else {
                var friends = document.getElementById('friends');
                response.data.forEach(function(item) {
                var d = document.createElement('div');
                d.innerHTML = item.name;
                data.appendChild(d);
                            });
            }
});
}
```

In the getFriendList() function, a call to Facebook API FB.api() is made. The first parameter is the path of the graph API provided by Facebook. In our example, '/me/friends' is used to get the friend list of the logged-in user. The second parameter is the JavaScript callback function that receives a response. The following operations are performed in the callback method:

1.   Check the response status whether its successful response using 'response.error'

2.   If it's a successful response, the result data available in response is iterated.

3.   'div' element is created for each item and appended to 'friends' div defined in index.html

Next, we will modify the login() function to call getFriendList() on a successful login.

```
function login() {
FB.login(
function(response) {
        if (response.session) {
            getFriendList();
        } else {
            alert('not logged in');
        }
            },
{ perms: "email" }
    );
}
```

Here we are checking the successful response of the 'response.session' value. If it is valid, we are calling the getFriendList() function.

Now the last step is to use the JavaScript functions with the PhoneGap's initialization event.

```
document.addEventListener('deviceready',
function() {
    try {
        /* Initialize the Facebook plug-in. Note that you need to replace the
        <app_id>by your Facebook's app_id */
        FB.init({ appId: "<app_id>", nativeInterface:PG.FB });
        document.getElementById('data').innerHTML = "";
        login();
    } catch (e) {
        alert(e);
    }
}, false);
```

Finally, to run the application, you need to put your Facebook's app_secret key into the AndroidManifest.xml file as shown in Figure 8–6.

Figure 8–6. *Facebook app_secret key*

You can get the Facebook app_id and app_secret from the Facebook developer site, https://developers.facebook.com/apps.

Run the FacebookPluginExample as an Android application on the simulator. The first screen will show the Facebook login page as shown in Figure 8–7.

Figure 8–7. *Facebook login screen*

After successful login, you will see your friend list as shown in Figure 8–8.

Figure 8–8. *Facebook friend list*

You can use jQueryMobile or Sencha Touch, along with the Facebook PhoneGap plug-in, to develop an attractive Facebook application. Also, you can call other Facebook graph API by using the Facebook plug-in to add more features.

C2DM Plug-In for Mobile Push Notification to PhoneGap

Push notification or server push is the latest way to send data from the server to the client. Have you noticed how Gmail receives and displays the new email that arrives in your inbox? You don't need to refresh the browser or click some refresh button to send request and receive latest data from the server.

In the recent past, polling was a popular technique to receive notification. The polling technique sends periodic requests to the server and refreshes the UI with the response received. You can think of it as a background process that sends requests at certain predefined intervals and receives the update or notification from the server. There are many known shortcomings to this approach. The major drawback of the polling approach is identifying the appropriate intervals to send the requests. With shorter intervals, there might be unnecessary requests and response trips that result in a loss of bandwidth and server resources. A larger interval might beat the purpose of polling because there might be a delay in receiving notification and it would no longer serve the purpose for which it was sent. If no new data is available, this approach consumes the battery of a mobile phone.

A server push allows the server to send the notification or update to the client without waiting for a request. In the push technique, the client doesn't have any background process to make the periodic request. At any point the server has an update, it can push the update to all registered clients. If the client is a mobile application, this technique is called a mobile push.

Your PhoneGap application can also leverage the mobile push technique through PhoneGap plug-in. Let's create a small PhoneGap application for an Android platform to receive push notifications from a C2DM service.

Setting Environment for Android

First, we will need to setup the PhoneGap project for Android. Refer to Chapter 2 to setup your project for Android. An Android project configuration is shown in Figure 8–9.

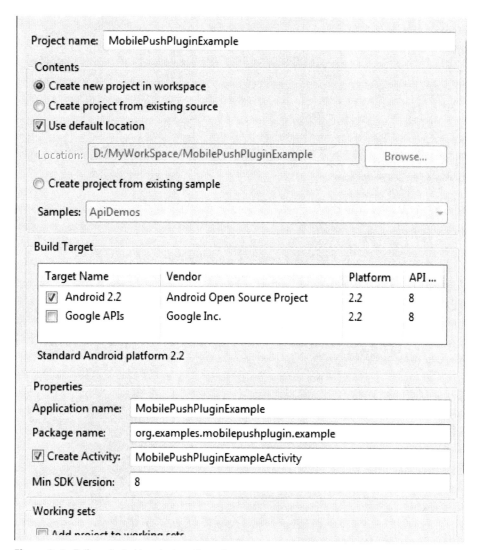

Figure 8–9. *Eclipse Android project configuration*

We will use an Android Cloud to Device Messaging (C2DM) framework for push notification. You can read more about C2DM service at http://code.google.com/android/c2dm/#intro.

Download the C2DM PhoneGap plug-in from http://github.com/awysocki/C2DM-PhoneGap/downloads. The C2DM-PhoneGap plug-in is a zip file. Unzip it in your favorite folder. The folder structure should look similar to the listing in Figure 8–10.

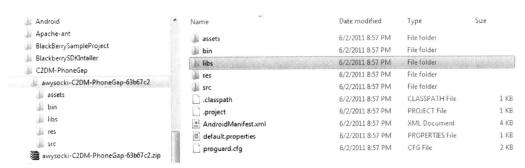

Figure 8–10. *C2DM-Plugin folder structure*

Next you will need to perform the following installation steps:

1. Register C2DMPlug-in

    Add the following XML element as a child of the "plugins" element in the plugins.xml file as shown in Figure 8–11. You may need to create an xml folder under the 'res' folder and copy plugins.xml from the PhoneGap Android sample application.

```
<plugin name="C2DMPlugin" value="com.plugin.C2DM.C2DMPlugin" />
```

Figure 8–11. *Plug-in registration*

2. Include the native part of the plug-in into the project

Copy the src folder from the C2DM-plug-in folder as shown in Figure 8–12 and paste it into the root of our PhoneGap application, i.e., "MobilePushPluginExample".

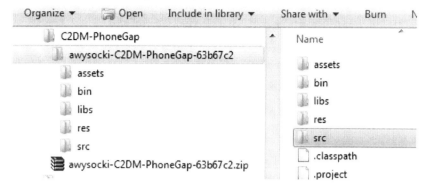

Figure 8–12. *C2DM-Plug-in native part*

3. Include the JavaScript part of the plug-in into the project

   Copy the following files from the C2DM-plug-in folder as shown in Figure 8–13 and paste them into the assets folder of our application.

   ▓ C2DMPlugin.js

   ▓ jquery_1.5.2.min.js

   ▓ PG_C2DM_script.js

   ▓ index.html

   Note that, even though index.html is not part of the plug-in, we are using it in our project to save time to create and include js files.

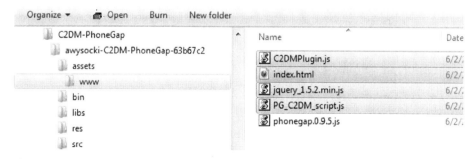

Figure 8–13. *JavaScript part of plug-in*

Note that we are not including the Phonegap.0.9.5.js file, as we are using PhoneGap-1.1.0. We will do the required modification for PhoneGap-1.1.0.

4. Finally we need to add that the required permissions in AndroidManifest.xml file for C2DM.Manifest file should be similar to the following listing:

```xml
<?xml version="1.0" encoding="utf-8"?>
<manifest xmlns:android="http://schemas.android.com/apk/res/android"
            package="org.examples.mobilepushplugin.example" android:versionCode="1"
            android:versionName="1.0">

        <uses-permission android:name="android.permission.CAMERA" />
        <uses-permission android:name="android.permission.VIBRATE" />
        <uses-permission android:name="android.permission.ACCESS_COARSE_LOCATION" />
        <uses-permission android:name="android.permission.ACCESS_FINE_LOCATION" />
        <uses-permission
android:name="android.permission.ACCESS_LOCATION_EXTRA_COMMANDS" />
        <uses-permission android:name="android.permission.READ_PHONE_STATE" />
        <uses-permission android:name="android.permission.INTERNET" />
        <uses-permission android:name="android.permission.RECEIVE_SMS" />
        <uses-permission android:name="android.permission.RECORD_AUDIO" />
        <uses-permission android:name="android.permission.MODIFY_AUDIO_SETTINGS" />
        <uses-permission android:name="android.permission.READ_CONTACTS" />
        <uses-permission android:name="android.permission.WRITE_CONTACTS" />
        <uses-permission android:name="android.permission.WRITE_EXTERNAL_STORAGE" />
        <uses-permission android:name="android.permission.ACCESS_NETWORK_STATE" />

        <!-- START:C2DM messaging stuff -->
        <uses-library android:name="com.google.android.c2dm.C2DMessaging" />

        <permission
android:name="org.examples.mobilepushplugin.example.permission.C2D_MESSAGE"
                    android:protectionLevel="signature" />
        <uses-permission
android:name="org.examples.mobilepushplugin.example.permission.C2D_MESSAGE" />

        <uses-permission android:name="com.google.android.c2dm.permission.RECEIVE" />

        <uses-permission android:name="android.permission.WAKE_LOCK" />
        <!-- END:C2DM messaging stuff -->
        <uses-sdkandroid:minSdkVersion="8" />

        <application android:icon="@drawable/icon" android:label="@string/app_name">
                <activity android:name=".MobilePushPluginExampleActivity"
                        android:label="@string/app_name">
                        <intent-filter>
                                <action
android:name="android.intent.action.MAIN" />
                                <category
android:name="android.intent.category.LAUNCHER" />
                        </intent-filter>
                </activity>
                <!-- START:C2DM messaging stuff -->
                <service android:name=".C2DMReceiver" />

                <!-- Only C2DM servers can send messages for the app. If
permission is
                        not set - any other app can generate it -->
                <receiver
android:name="com.google.android.c2dm.C2DMBroadcastReceiver"

android:permission="com.google.android.c2dm.permission.SEND">
                        <!-- Receive the actual message -->
```

```
                                <intent-filter>
                                        <action
android:name="com.google.android.c2dm.intent.RECEIVE" />
                                        <category
android:name="org.examples.mobilepushplugin.example" />
                                </intent-filter>
                                <!-- Receive the registration id -->
                                <intent-filter>
                                        <action
android:name="com.google.android.c2dm.intent.REGISTRATION" />
                                        <category
android:name="org.examples.mobilepushplugin.example" />
                                </intent-filter>
                        </receiver>
                        <!-- END:C2DM messaging stuff -->
                </application>

</manifest>
```

Once you are done with the previous three steps, you will see the MobilePushPluginExample project structure as shown in Figure 8–14.

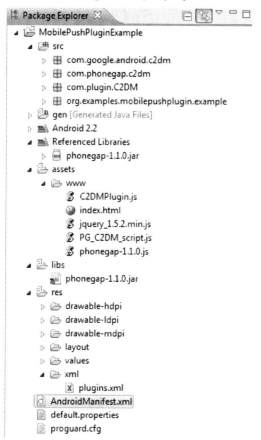

Figure 8–14. *MobilePushPluginExample project structure*

Modifying Plug-In for PhoneGap-1.1.0

At the time of the writing this book, a C2DM plug-in is based on the PhoneGap version 0.9.5. To use it with PhoneGap-1.1.0, we have to complete a couple of modifications.

1. `C2DMPlugin.jsfiles`.

    Open C2DMPlugin.js and go to following function definition:

```
PhoneGap.addConstructor(function() {
        //Register the javascript plugin with PhoneGap
        PhoneGap.addPlugin('C2DM', new C2DM());

        //Register the native class of plugin with PhoneGap
        PluginManager.addService("C2DMPlugin",
"com.plugin.C2DM.C2DMPlugin");

        //alert( "added Service C2DMPlugin");
});
```

   Remove the following line:

```
//Register the native class of plugin with PhoneGap
PluginManager.addService("C2DMPlugin","com.plugin.C2DM.C2DMPlugin");
```

   In PhoneGap-1.1.0, the plug-in has to be registered in the plugin.xml file. PluginManager is no longer available in the PhoneGap-1.1.0. We have already registered the C2DM-plugin in the plugin.xml file.

   Now the modified function looks as follows:

```
PhoneGap.addConstructor(function() {
        //Register the javascript plugin with PhoneGap
        PhoneGap.addPlugin('C2DM', new C2DM());

        //alert( "added Service C2DMPlugin");
});
```

2. Move `C2DMReceiver.java`

    We have to move C2DMReceiver.java from com.phonegap.c2dm package to our application project, i.e., org.examples.mobilepushplugin.example. To do so, drag the C2DMReceiver.java from com.phonegap.c2dm package and drop it into org.examples.mobilepushplugin.example, as shown in Figure 8–15.

Figure 8–15. *C2DMReceiver.java location*

We also have to modify the PhoneGap version in the index.html file, as we copied it from the plug-in folder. To do so, replace phonegap.0.9.5.js by phonegap-1.1.0 in the script tag.

Signup for C2DM Service

Go to http://code.google.com/android/c2dm/signup and fill in the form to register a sender. You have to mention the package name of the Android application and the sender account email along with other information. For our application, the package name should be "org.examples.mobilepushplugin.example" and we should use a Google account as a sender account email. We have to use the sender account email id in our PhoneGap application to register the device for receiving notification.

Using C2DM Sender Account in PhoneGap

The C2DM plug-in comes with ready-to-use PhoneGap deviceready implementation. We have to use our C2DM sender account to register the device for notification.

Open the PG_C2DM_script.js file and go to the PhoneGap deviceready event implementation. Modify the "your_c2dm_account@gmail.com" to our C2DM sender account as shown in Figure 8–16.

Figure 8–16. *Registering application with C2DM sender account*

Android Simulator for C2DM-Enabled Service

You need to use an AVD (Android Virtual Device) with target "Google APIs(Google Inc.) - API Level 8" to run the C2DM-enabled Android application. Please refer to Chapter 2 to create new AVD from "Android SDK and AVD manager".

You also have to add your Google account in the simulator. To do so, run the simulator and open the setting as shown in Figure 8–17.

Figure 8–17. *Android setting option*

Go to "Account & Sync" and click on Add Account. You have to enter your Google account Id and password. Note that it's not the C2DM sender account. It's your Google account that you use to retrieve your email and other stuff from your Android phone.

Now you are all set to test the C2DM plug-in. Run the MobilePushPluginExample as an Android application. Make sure the target simulator is Google API's simulator. You will see the screen shown in Figure 8–18 on the simulator.

Figure 8–18. *MobilePushPluginExampleOutput on simulator*

To understand the output appearing on the screen, we will go through the devicereadyevent callback function in the PG_C2DM_script.js file.

Here, the window.plugins.C2DM.register() is calling the plug-in's method to register the device or simulator into C2DM service. On successful registration, the C2DM server returns the Registration Id (REGID). This REGID is used to push the notification message. But wait, our device is not supposed to push the message. It's a notification receiver right? Here, we have to understand the role of the application server between our PhoneGap application running on the mobile and the C2DM service hosted on Google.

Let's use an example to understand how C2DM push works. Assume that MobilePushPluginExample is installed on multiple Android phones. Now, each phone receives the REGID from a C2DM service. Essentially, each REGID is unique. Before sending the REGID, the C2DM service stores all of the required information about the device and the network for further use in sending notification. The C2DM is the one that sends the notification to devices. Now, it's the responsibility of the intermediate application server to identify the updated data and ask C2DM to send notification to an actual mobile. To do so, our server has to know the REGIDs.

Usually, a C2DM-enabled mobile application sends the REGID to our server. The server stores the REGIDs for all mobiles that are running this application. Once the server decides to send the notification, it uses the REGID to ask the C2DM service to do so.

We received the REGID from the C2DM service, as can be seen in Figure 8–18. Now, you can send the push notification by using this REGID. You can use Java servlet or php to create a server side code to send the message. To learn more about how to send push notification go to the Android C2DM site:
http://code.google.com/android/c2dm/index.html#push.

In addition, there is a command line tool available to simulate the server. Go to http://curl.haxx.se/download.html and download the platform specific curl tool. There are two steps to sending the notification:

1. Get the authentication key

   Run the following command on the console:

```
D:\cURL>curl https://www.google.com/accounts/ClientLogin -d Email=<C2DM Sender Account>-
d "Passwd=<password>" -d accountType=GOOGLE -d source=org.examples.
mobilepushplugin.example -d service=ac2dm –k
```

   You have to replace <C2DM Sender Account> and <password> with your registered Google account and password.

   After running the above command, you will get an authentication key similar to this listing:

```
Auth=DQAAAMEAAABrqkqH2KYjDfCD93tndEF7n81lKgf5vczCwELPSXgW6xm_9EACDuOlsJFGud7fNBI
HcRV1Q6zUmLwxFFJqosdn1nYYmGahOyu7fpT8vfjNLAVx8hs5aymz9OULg-pzKOyWWa1-6BDci1TBCoP
2q6ZwJqEjzH6rArHSlD9DhruEKBrogjfBAWyeIm2fs9THvEkilSMO2Q8utoqyfGOid9keCQad5QPV7oO
vNSe6urKOV4ZWEKxG7KAlXCsjW18u_m2Az6jj7DlUoVD89MeLvXOW
```

2. Send the message to your application running on a simulator

   To send the notification, we will use the authentication key and REGID by using the following curl command:

```
S:\cURL>curl --header "Authorization: GoogleLoginauth=<Auth Key>"
"https://android.apis.google.com/c2dm/send" -d registration_id=<REGID>-d
"data.message=This is a test message" -d "data.msgcnt=1" -d collapse_key=0 –k
```

   Replace <Auth Key> with the authentication key received from step 1 and REGID received by the application on the simulator. We are sending "This is a test message" text as a push notification to the simulator.

We will see the notification received by our application on the simulator as shown in Figure 8–19.

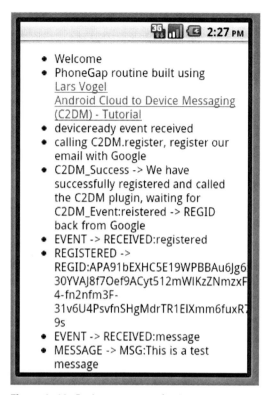

Figure 8–19. *Push message on simulator*

If you want to leverage the push notification service for iPhone-PhoneGap application, you can use the PhoneGap plug-in available at: `https://github.com/urbanairship/ios-phonegap-plugin`. It uses an Urban Airship service for push mobile notification.

Conclusion

PhoneGap plug-ins dynamically extend the PhoneGap application to include out of bound features. A PhoneGap application can use approximately any native features by using plug-ins.

Plug-ins are good friends of PhoneGap, but the community support for plug-ins is still at an early stage. At the same time, the organization behind PhoneGap is making popular plug-ins official. However, plug-in support is still far from perfect. One example of this is trying to write a Facebook Connect application for the iPhone. When we tried to use this plug-in with PhoneGap 1.1.0, we found that it did not work. We also found that including this plug-in was very cumbersome. Our guess is that for upcoming releases of PhoneGap, the support for Plug-ins will improve and they will be much easier to bundle and use in a PhoneGap Application.

In this chapter, we only talked about the Android PhoneGap plug-in for Facebook Connect and Cloud Push. We added pointers about the iPhone plug-ins for the same reason, which need to improve before it can be easily and effectively used.

Chapter 9

Extending PhoneGap

Thus far, we have seen that PhoneGap has two parts

1. The JavaScript part that we call from our PhoneGap applications

2. A native part we include in our PhoneGap project to expose native phone features.

These two parts work for scenarios where we want to access common phone features, including the following:

1. Camera

2. Accelerometer

3. File system

4. Geo location

5. Storage services

However, we often need to go beyond these features.

JavaScript Limitations

We have seen that JavaScript has improved in performance in the last decade; it has become 100 times faster than it was five years ago. However, even when this is true, sometimes applications need to do heavy lighting, do things in the background, or do complex operations. These are best done in native code for performance reasons.

For example, if we want to download a multipart file, it involves downloading different parts of the file parallel and then checking its checksum. This part is best done in Java for Android and in Objective-C for iPhone.

Solution

If you recall from Chapter 1, we said PhoneGap is a bridge between the JavaScript world and the native world. The entire PhoneGap framework is based on plug-in architecture. This means PhoneGap provides a mechanism by which we map JavaScript functions (and arguments, return types, and callbacks) to native code.

We can add a native code to the PhoneGap application and expose the code easily using JavaScript. For this, we need two parts

1. Native code that does heavy lifting

2. JavaScript code that exposes this native code

Both are glued by the PhoneGap framework.

Architecture

The PhoneGap architecture is shown in Figure 9–1. As we observed, PhoneGap has two parts: the PhoneGap JavaScript engine and the PhoneGap native engine. We add the native code as a plug-in to the PhoneGap native engine and add JavaScript code as a plug-in to the PhoneGap JavaScript engine.

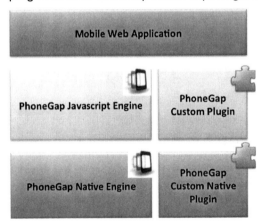

Figure 9–1. *PhoneGap architecture*

Scope

This chapter focuses on how you can extend PhoneGap functionality to expose more of your native code.

However, please note that even if you write PhoneGap plug-ins, the only way to inject the plug-in is to add the plug-in source to your project. There is currently no way to build a plug-in into a package and add the package to your PhoneGap project. This stops you using your custom plug-ins when you are using PhoneGap build.

For this chapter, let's keep the nature of the plug-in very simple. We call this a helloworld plug-in. We pass a name to the plug-in, and we get back a string "Hello <name>! The time now is <Current Time>".

This way, we focus mainly on the bridge aspect of the plug-in.

Extending PhoneGap for Android

To begin, we create the plug-in as part of an Android PhoneGap application and then extract the plug-in out. This is required because

1.   The plug-in requires PhoneGap jar.

2.   We need to test the plug-in.

A plug-in has two parts, one on each side of the PhoneGap framework (bridge). We have a native part (a class-extending plug-in) and a JavaScript file using PhoneGap's JavaScript framework.

Before we begin, let's create an Android PhoneGap project (see Figure 9–2). This is shown in Chapter 2.

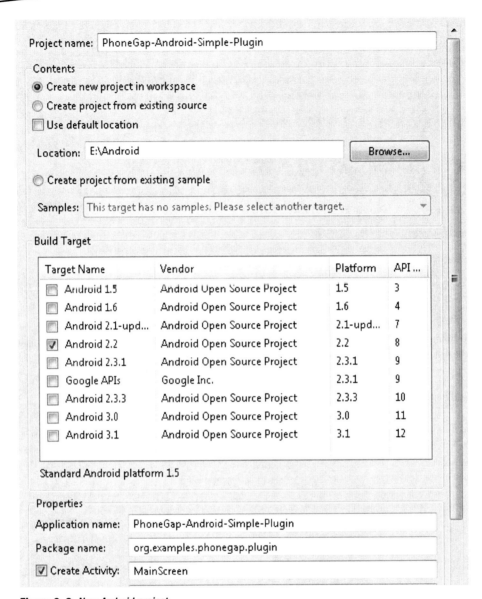

Figure 9–2. *New Android project*

Then we need to configure the base Android project for PhoneGap.

1. Change the MainScreen class to extend DroidGap.

2. Add PhoneGap jar to classpath.

3. Add the PhoneGap JavaScript library to the assets/www folder.

The Android project looks as shown in Figure 9–3.

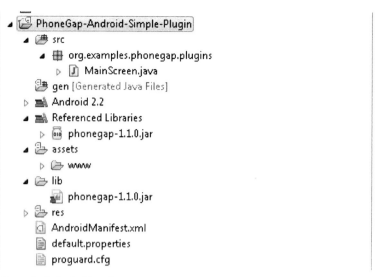

Figure 9–3. *Android project structure*

Declaring the Native Part of the Plug-In

Now we add an appropriate package for the plug-in, say
"org.examples.phonegap..plugins.simpleplugin." Then we declare a class named
Simple Plug-in, which extends PhoneGap's com.phonegap.api.Plugin class, as shown in
Figure 9–4.

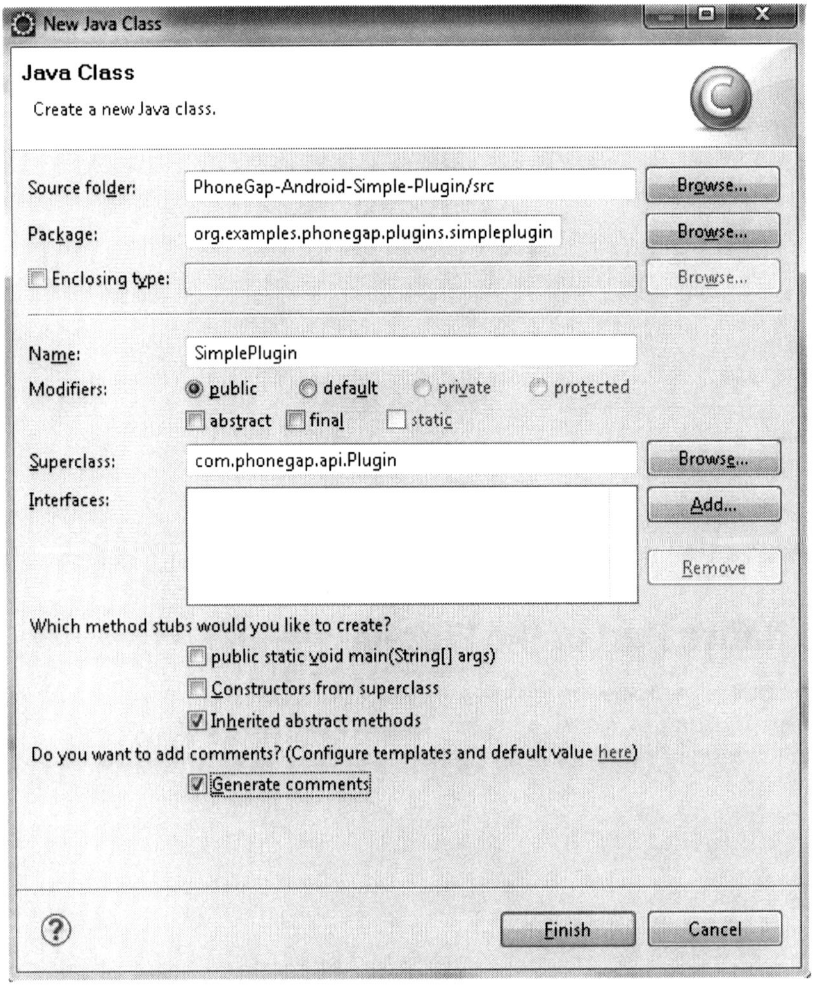

Figure 9–4. *Declaring the native part of the plug-in*

Once you click on the "Finish" button, you would get the code shown below

```
package org.examples.phonegap.plugins.simpleplugin;

import org.json.JSONArray;

import com.phonegap.api.Plugin;
import com.phonegap.api.PluginResult;

/**
 * @author rohit
 *
 */
public class SimplePlugin extends Plugin {

        /* (non-Javadoc)
```

```
         * @see com.phonegap.api.Plugin#execute(java.lang.String, org.json.JSONArray,
java.lang.String)
         */
        @Override
        public PluginResult execute(String action, JSONArray data, String callbackId) {
                // TODO Auto-generated method stub
                return null;
        }

}
```

When we extend the `com.phonegap.api.Plugin` class, we have to implement the execute method. The arguments of the execute method are

1. *Action*: The action to be performed. For example, for a file-based plug-in, could be open, close, read, write, etc.

2. *Data*: The data passed from the JavaScript side of the plug-in. This is the data passed from the PhoneGap's JavaScript app to the native code. For example, for a file-based plug-in, could be filename, data, etc.

3. *CallbackId*: This is used when calling back the JavaScript function.

The return type of the execute method is PluginResult. PluginResult typically takes a Status enum and one other argument depicting the cause or more information.

For example, new PluginResult(Status.OK);

Status enum has many values; all are depicted below (the names are self-explanatory)

1. NO_RESULT

2. OK

3. CLASS_NOT_FOUND_EXCEPTION

4. ILLEGAL_ACCESS_EXCEPTION

5. INSTANTIATION_EXCEPTION

6. MALFORMED_URL_EXCEPTION

7. IO_EXCEPTION

8. INVALID_ACTION

9. JSON_EXCEPTION

10. ERROR

Following is the implementation of the hello plug-in, which takes a name and returns the "Hello <name>! The time is <time>" text.

```java
package org.examples.phonegap.plugins.simpleplugin;

import java.util.Date;

import org.json.JSONArray;
import org.json.JSONException;

import com.phonegap.api.Plugin;
import com.phonegap.api.PluginResult;
import com.phonegap.api.PluginResult.Status;

/**
 * @author rohit
 *
 */
public class SimplePlugin extends Plugin {

        public static String ACTION_HELLO="hello";

        /*
         * (non-Javadoc)
         *
         * @see com.phonegap.api.Plugin#execute(java.lang.String,
         * org.json.JSONArray, java.lang.String)
         */
        @Override
        public PluginResult execute(String action, JSONArray data, String callbackId) {
                PluginResult pluginResult = null;
                if (ACTION_HELLO.equals(action)) {

                        String name = null;
                        try {
                                name = data.getString(0);

                                String result = "Hello " + name + "! The time is "
                                                + (new Date()).toString();

                                pluginResult = new PluginResult(Status.OK, result);

                                return pluginResult;
                        } catch (JSONException e) {
                                pluginResult = new PluginResult(Status.JSON_EXCEPTION,
"missing argument name");
                        }
                } else {
                        pluginResult = new PluginResult(Status.INVALID_ACTION,
                                        "Allowed actions is hello");
                }
                return pluginResult;
        }

}
```

You can see in the above code that we explicitly check for an action before we process
the request. If the action is not what is handled by the plug-in, we return
`Status.INVALID_ACTION`. The second check is for the argument. If we get any JSON
exception while fetching the first argument as a string, we return `Status.INVALID_JSON`.

When the action and the argument are correct, we create a string "Hello <name>! The time is <time>" and return it with `Status.OK`.

Please note, you do not have to spawn any threads from this method. Your entire method can be synchronous. This will not hand the JavaScript plug-in call calling this code. This is internally handled by PhoneGap, and that's why we have success and failure callback in JavaScript (which you will see in the following section).

Declaring the JavaScript Part of the Plug-In

The JavaScript part of this plug-in is declared in a file named simpleplugin.js. There are three steps to declare the JavaScript part of plug-in:

1. Plug-in Registration

   In the JavaScript part of the PhoneGap plug-in, things begin from the call to add the plug-in in PhoneGap.

```
PhoneGap.addConstructor(function() {
            // Register the Javascript plug-in with PhoneGap
            PhoneGap.addPlugin('SimplePlugin', new SimplePlugin());
});
```

   The plug-in is registered in the /res/xml/plugins.xml file. Add the following XML element as a child of the "plugins" element in the plugins.xml file:

```
<plugin name="SimplePlugin"
value="org.examples.phonegap.plugins.simpleplugin.SimplePlugin" />
```

   Note: here we are doing two things

   a. Registering a JavaScript object as a plug-in with the name "SimplePlugin."

   b. Registering a PhoneGap Java class as a service named "SimplePlugin." You can think of this as an alias for the class name "`org.examples.phonegap.plugins.simpleplugin.Simple Plugin`."

2. Create the JavaScript object SimplePlugin.

   This is done by declaring a JavaScript function.

```
var SimplePlugin = function() {
}
```

3. Add a plug-in function.

   In this step, we will add the plug-in function, which our JavaScript will call. In the following function, we are actually delegating the call to the native PhoneGap bridge asking it to actually call out the "`SimplePlugin`" service, which is the "`org.examples.phonegap.plugins.simpleplugin.SimplePlugin`" class. Furthermore, we are registering two callbacks: a callback if the call is successful and one when the call fails. Then we are declaring the action we want to invoke.

You can recall, we have code in our plug-in class to handle the "hello" service. Last, remember our plug-in class's execute method takes an argument JSONArray; here, we are passing it as [name].

```
SimplePlugin.prototype.hello = function(name, successCallback, failureCallback) {

        PhoneGap.exec(
successCallback, // Success Callback
failureCallback, // Failure Callback
'SimplePlugin',  // Registered plug-in name
'hello', // Action
[name] //Argument passed in
);
      };
```

The complete JavaScript file simpleplugin.js is as follows:

```
/**
 *
 * @return Instance of SimplePlugin
 */
var SimplePlugin = function() {

}

/**
 * @param name
 *            The name passed in
 * @paramsuccessCallback
 *            The callback that will be called when simple plugin runs
 *            successfully
 * @paramfailureCallback
 *            The callback that will be called when simple plugin
 *            fails
 */
SimplePlugin.prototype.hello = function(name, successCallback, failureCallback) {
    PhoneGap.exec(successCallback, // Success Callback
                failureCallback, // Failure Callback
                'SimplePlugin',  // Registered Plug-in name
                'hello',                 // Action
                [ name ]);          // Argument passed in
};

/**
 * <ul>
 * <li>Register the Simple Listing Javascript plugin.</li>
 * </ul>
 */
PhoneGap.addConstructor(function() {
        // Register the Javascript plug-in with PhoneGap
        PhoneGap.addPlugin('SimplePlugin', new SimplePlugin());
});
```

Calling the Plug-In

Time to test our plug-in. To do so, we need the following:

1. HTML file

2. PhoneGap js file

3. Plug-in js file

4. Plug-in Java file

Your Android project should look as shown in Figure 9–5.

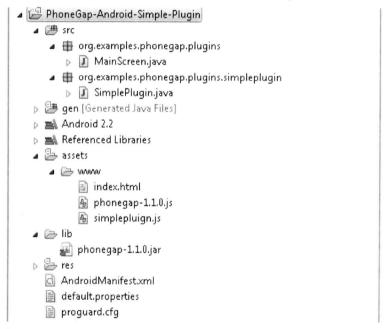

Figure 9–5. *Android PhoneGap plug-in project structure*

Your index.html file should look as follows:

```
<!DOCTYPE HTML>
<html>

    <head>
        <title>PhoneGap</title>
        <script type="text/javascript" charset="utf-8" src="phonegap-1.1.0.js"></script>
        script type="text/javascript" charset="utf-8" src="simpleplugin.js"></script>
        <script type="text/javascript" charset="utf-8">
        document.addEventListener('deviceready', function() {
            var btn = document.getElementById("hello");
            var textbox = document.getElementById("name");
            var output = document.getElementById("output");
```

```
                btn.addEventListener('click', function() {

                    var text = textbox.value;

                    window.plugins.SimplePlugin.hello(text,
                    //success callback


                    function(result) {
                        output.innerHTML = result;
                    }
                    //failure callback,
                    , function(err) {
                        output.innerHTML = "Failed to invoke simple plugin";
                    });
                });

        }, true);
        </script>
    </head>

    <body>
        <h1>
            Simple Plugin Demo
        </h1>
        <table border="1">
            <tr>
                <td>
                    Enter Name
                </td>
                <td>
                    <input type="text" name="name" id="name">
                    </input>
                </td>
            </tr>
            <tr>
                <td>
                    <b>
                        Output:
                    </b>
                </td>
                <td>
                    <div id="output">
                    </div>
                </td>
            </tr>
            <tr>
                <td colspan="2">
                    <button id="hello">
                        Say Hello
                    </button>
                </td>
            </tr>
        </table>
    </body>

</html>
```

Here you should notice that the plug-in is invoked as follows. We first pass the text containing the name, and then we register a callback for success and one for failure.

```
window.plugins.SimplePlugin.hello(
    text,
    //success callback
    function (result) {
        output.innerHTML = result;
    },
    //failure callback
    function (err) {
        output.innerHTML = "Failed to invoke simple plugin";
    }
);
```

Finally, when we run this Android project, we see the following output as shown in Figure 9–6.

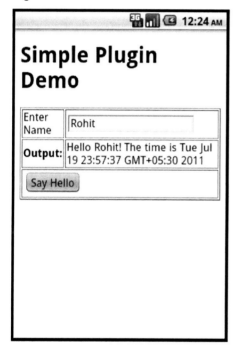

Figure 9–6. *PhoneGap plug-in output on Android*

Sharing the Android PhoneGap Plug-In

As far as PhoneGap framework version 1.1.0 is concerned (the version at the time this book was written), there is no way to package and share your plug-in.

The only way to share your plug-in is by

1.   Sharing the Java source file

    2. Sharing the JavaScript source file

    3. Readme file telling how the plug-in is to be used

PhoneGap plug-ins are typically uploaded at `https://github.com/phonegap/phonegap-plugins`. If you wish to contribute your work, you can work with the PhoneGap team to add your plug-in in this repository.

Extending PhoneGap for iPhone

PhoneGap provides plug-ins for XCode for creating PhoneGap-based applications. At the time this book was written, PhoneGap moved from version 0.9.5 to 1.1.0. There are some changes in the iPhone-PhoneGap plug-in framework. This chapter focuses on 1.1.0 plug-in development.

Steps for installing the 1.1.0 XCode extension:

    1. Download PhoneGap 1.1.0 zip and unzip it.

    2. Go to the iOS folder and install PhoneGapInstaller.pkg.

Once you have installed the 1.1.0 XCode plug-in for PhoneGap, create a PhoneGap-based application from XCode as depicted in Figure 9–7 and Figure 9–8.

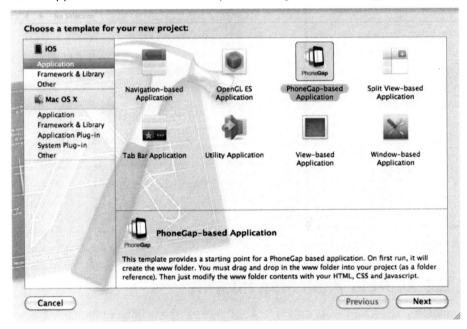

Figure 9–7. *Create a new iOS PhoneGap project*

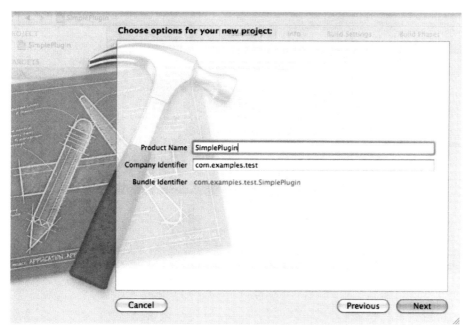

Figure 9–8. *Create a new iOS PhoneGap project*

Follow the steps in Chapter 3 to add the www folder to the project. Now run the project and ensure that you are able to see the iPhone PhoneGap-based application.

Declaring the Native Part of the Plug-In

The native part of the plug-in for PhoneGap 1.1.0 needs to be added to the plug-ins folder. This is depicted in Figure 9–9.

Figure 9–9. *iPhone plug-in native part*

Create an Objective-C class to the plug-ins folder. Let's name the class SimplePlugin. SimplePlugin extends the PGPlugin. The SimplePlugin.h file looks as follows.

```
#import <Foundation/Foundation.h>
#ifdef PHONEGAP_FRAMEWORK
#import <PhoneGap/PGPlugin.h>
#else
#import "PGPlugin.h"
#endif

@interface SimplePlugin :PGPlugin {

}
/**
 * Sets the idleTimerDisable property to true so that the idle timeout is disabled
 */
- (void) hello:(NSMutableArray*)arguments withDict:(NSMutableDictionary*)options;


@end
```

Here we declare a function name "hello," which has the following signature:

```
- (void) hello:(NSMutableArray*)arguments withDict:(NSMutableDictionary*)options;
```

This function does not return anything. Instead, it takes two arguments

1. Arguments

2. Options

Any arguments to the plug-in or an input (in our case the name) is passed using the "arguments."

Now let's implement the SimplePlugin's hello function. In the first version, we will return a string "hello world" from the plug-in. In addition, we will explain how to gain access to the arguments passed and how to call success and failure callbacks.

Note the plug-in is called from JavaScript as follows:

```
window.plugins.SimplePlugin.hello(
        "Bob",
//success callback
function(result){
                alert("plugin returned "+result);
},
//failure callback,
function(err){
                alert("got error when invoking the plugin");
        }
);
```

Following is the skeleton code of the plug-in method.

The plug-in can gain access to the input argument, in our case "Bob," extracting it from the arguments object. Note the first object in the arguments array is always the callbackId, used to call back the JavaScript callback functions. We can extract the actual arguments (in our case, only "Bob") from index 1 onwards.

```
NSString * name = [arguments objectAtIndex:1];
```

If we have another argument, we would access it at index 2.

Now let's concentrate on how to invoke either the success or failure JavaScript callback functions. This begins with the declaration of the PluginResult object. This is followed by declaring two more objects, one for callbackId (which helps us call the callback functions) and another is a string, the JavaScript string, that we will embed in the HTML page to actually call the callbacks.

```
NSString* jsString = nil;
NSString* callbackId = [arguments objectAtIndex:0];
```

Now let's go through the flow of code for success and failure conditions. This is shown in the code below.

If things are going fine, we create a result object with status PGCommandStatus_OK. Then we go and create the jsString object from the result, passing the callbackId. Finally, we write the JavaScript to actually call the success callback by calling [self writeJavascript:jsString].

In the case of a failure, we create a PluginResult object with status other than PGCommandStatus_OK and create the jsString for error/failure callback. Finally, we invoke the error/failure callback using [self:writeJavascript:jsString]:

```
PluginResult* result=nil;
NSString* jsString=nil;
NSString* callbackId=[argumentsobjectAtIndex:0];


if(success){
result=[PluginResultresultWithStatus:PGCommandStatus_OK];
    jsString=[resulttoSuccessCallbackString:callbackId];
}
else{
result=[PluginResultresultWithStatus:PGCommandStatus_ILLEGAL_ACCESS_EXCEPTION];
    jsString=[resulttoErrorCallbackString:callbackId];
}

[selfwriteJavascript:jsString];
```

If we want to pass data when we are calling the success or failure callback, we can do so by passing an addition argument when creating the PluginResult object. Here we pass a string by calling the resultWithStatus: messageAsString function of PluginResult.

```
result = [PluginResultresultWithStatus:PGCommandStatus_OK messageAsString:@"Hello
World"];
```

The complete SimplePlugin looks as follows. Note we do not have negative paths here, and therefore, we create only jsString for the success callback.

```
#import "SimplePlugin.h"


@implementation SimplePlugin
- (void) hello:(NSMutableArray*)arguments withDict:(NSMutableDictionary*)options
```

```
{
    PluginResult* result = nil;
    NSString* jsString = nil;
    NSString* callbackId = [arguments objectAtIndex:0];
    NSString* name = [arguments objectAtIndex:1];
    NSDate* date = [NSDate date];
    NSDateFormatter* formatter = [[[NSDateFormatteralloc] init] autorelease];

    //Set the required date format

    [formatter setDateFormat:@"yyyy-MM-ddhh:mm:ss"];

    //Get the string date

    NSString* dateStr = [formatterstringFromDate:date];


    NSString* returnStr = [NSStringstringWithFormat:@"Hello %@.The time is  %@!",
name,dateStr];

    result = [PluginResultresultWithStatus:PGCommandStatus_OK
messageAsString:returnStr];
    jsString = [result toSuccessCallbackString:callbackId ];

    [selfwriteJavascript:jsString];
}
@end
```

Just creating the .h and .m files for the plug-ins and putting the files in the plug-ins folder is not enough. We need to register our SimplePlugin with the PhoneGap framework. Adding an entry to the PhoneGap.plist file in the Supporting Files folder does this.

This is shown in Figure 9–10.

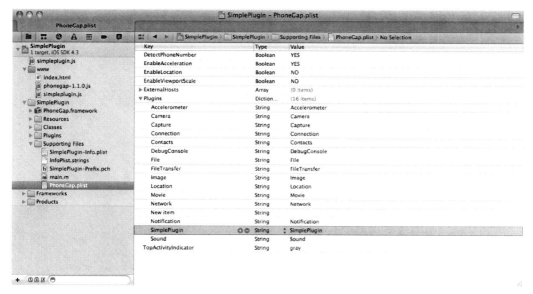

Figure 9–10. *Register PhoneGap plug-in*

Declaring the JavaScript Part of the Plug-In

The JavaScript part of the iPhone plug-in is different from what you have seen for Android.

This is done in mainly two steps

1. Declare a JavaScript class named SimplePlugin and add a method, in our case "hello," to it. In the hello function, we map the JavaScript arguments to the Objective Plugin class and method.

2. The second part is to create a method install for SimplePlugin and register the JavaScript plug-in by calling PhoneGap.addConstructor(SimplePlugin.install);

Let's focus on the hello function of the plug-in for a while. Note that we are calling the PhoneGap.exec function inside the plug-in.

Following the signature of PhoneGap.exec

```
PhoneGap.exec(<<successCallback>>,<<failureCallback>>,<<Plugin Name>>,<<Action
Name>>,<<Arguments Array>>)
```

Note how we pass the first argument of the hello function "name" as part of the arguments array. The successCallback and errorCallback go as the first and second arguments to the PhoneGap.exec function. The plug-in class and the method name go as the third and fourth arguments.

```
SimplePlugin.prototype.hello = function(name,successCallback, errorCallback) {
    PhoneGap.exec(
        successCallback,
        errorCallback,
```

```
        "SimplePlugin",
        "hello",
        [name]);
};
```

The complete code for the JavaScript part is shown below.

```
if (!PhoneGap.hasResource("simpleplugin")) {
    PhoneGap.addResource("simpleplugin");

    /**
     * @returns instance of powermanagement
     */

    function SimplePlugin() {};

    /**
     *
     * @param name Given the name, successCallBack gets the string "Hello <name>! The
time is <time>."
     * @paramsuccessCallback function to he called when the wake-lock was acquired
successfully
     * @paramerrorCallback function to be called when there was a problem with acquiring
the wake-lock
     */
    SimplePlugin.prototype.hello = function (name, successCallback, errorCallback) {
        PhoneGap.exec(successCallback, errorCallback, "SimplePlugin", "hello", [name]);
    };


    /**
     * Register the plug-in with PhoneGap
     */
    SimplePlugin.install = function () {
        if (!window.plugins) window.plugins = {};

        window.plugins.SimplePlugin = new SimplePlugin();

        return window.plugins.SimplePlugin;
    };

    PhoneGap.addConstructor(SimplePlugin.install);
}
```

Calling the Plug-In

To test the plug-in, we will create a PhoneGap application and call the plug-in from
there. This part is exactly the same as that for Android.

You need to follow these steps

1. Include the PhoneGap 1.1.0 js file.

2. Include the simpleplugin.js file.

3. Register a button click to invoke the plug-in.

4. Register success and failure callback to show the result.

The complete source code of the index.html is as follows:

```
<!DOCTYPE HTML>
<html>

    <head>
        <title>PhoneGap</title>
        <script type="text/javascript" charset="utf-8" src="phonegap-1.1.0.js"></script>
        <script type="text/javascript" charset="utf-8" src="simpleplugin.js"></script>
        <script type="text/javascript" charset="utf-8">
        document.addEventListener('deviceready', function() {
            var btn = document.getElementById("hello");
            var textbox = document.getElementById("name");
            var output = document.getElementById("output");

            btn.addEventListener('click', function() {

                var text = textbox.value;

                window.plugins.SimplePlugin.hello(text,
                //success callback


                function(result) {
                    output.innerHTML = result;
                }
                //failure callback,
                , function(err) {
                    output.innerHTML = "Failed to invoke simple plugin";
                });
            });

        }, true);



        </script>
    </head>

    <body>
        <h1>
            Simple Plugin Demo
        </h1>
        <table border="1">
            <tr>
                <td>
                    Enter Name
                </td>
                <td>
                    <input type="text" name="name" id="name">
                    </input>
                </td>
            </tr>
```

```
        <tr>
            <td>
                <b>
                    Output:
                </b>
            </td>
            <td>
                <div id="output">
                </div>
            </td>
        </tr>
        <tr>
            <td colspan="2">
                <button id="hello">
                    Say Hello
                </button>
            </td>
        </tr>
    </table>
</body>

</html>
```

When you run the PhoneGap example, you will see the application as shown in Figure 9–11.

Figure 9–11. *PhoneGap plug-in output*

Sharing the iPhone PhoneGap Plug-In

You need to share the following files to share the plug-in.

1. SimplePlugin.h

2. SimplePlugin.m

3. simpleplugin.js

Add the above list of files to the Plugin documentation. Also document on how to invoke the plug-in from JavaScript.

Extending PhoneGap for BlackBerry

Similar to Android, PhoneGap's plug-in for BlackBerry has two parts, one on each side of the PhoneGap framework (bridge). We have a native part (a class-extending PhoneGap's plug-in) and a JavaScript file using PhoneGap's JavaScript framework.

We assume you are using BlackBerry WebWorks SDK version greater than version 1.5.

We assume that BlackBerry WebWorks SDK is installed on C:\BBWP, and we have Java 1.6 SDK and Ant installed and in path. We also assume our development directory is D:\PhoneGap-Plugin, and we have PhoneGap SDK present in D:\PhoneGap-plugin\PhoneGap-1.1.0 directory. Refer to Chapter 3 to recall the system requirement for BlackBerry PhoneGap development.

The steps for creating and testing BlackBerry plug-in are as follows:

1. Create the plug-in Java file and dump it inside PhoneGap SDK's framework.

2. Create the BlackBerry PhoneGap project to test the plug-in.

3. Then compile the BlackBerry PhoneGap project to compile the plug-in Java, which you dumped inside PhoneGap SDK's framework folder. If there is a compilation error, you need to delete the project you created in step 2, fix the Java file, and repeat from step 2.

4. When the plug-in Java file compiles without an error, then you can dump the JavaScript plug-in file and code the HTML page to use the plug-in.

Declaring the Native Part of the Plug-In

The BlackBerry plug-in class is very similar to the Android plug-in class. The only difference is the BlackBerry plug-in class uses the PhoneGap 1.1.0 framework; therefore, there are a few differences.

Following is the skeleton of the BlackBerry plug-in class:

```
package com.phonegap.plugins;

import com.phonegap.api.Plugin;
import com.phonegap.api.PluginResult;


import java.util.Date;
import com.phonegap.json4j.JSONArray;

public class HelloWorldPlugin extends Plugin {

    private static final String ACTION_HELLO="hello";

    /**
     * Executes the requested action and returns a PluginResult.
```

```
 *
 * @param action     The action to execute.
 * @paramcallbackIdThe callback ID to be invoked upon action completion.
 * @paramargsJSONArry of arguments for the action.
 * @return           A PluginResult object with a status and message.
 */
public PluginResult execute(String action, JSONArray data, String callbackId) {
        return null;
}

/**
 * Called when the plug-in is paused.
 */
public void onPause() {

}

/**
 * Called when the plug-in is resumed.
 */
public void onResume() {

}

/**
 * Called when the plug-in is destroyed.
 */
public void onDestroy() {

}

}
```

Note that we dump this plug-in class not in our project area but inside the PhoneGap
SDK area. The screenshot in Figure 9–12 shows where we copy this plug-in class. You
may have to create the plug-ins folder.

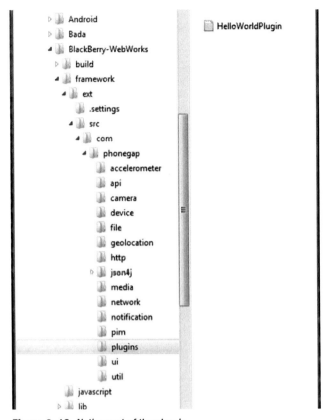

Figure 9–12. *Native part of the plug-in*

The main catch here is that you need a proper Java plug-in class in the above directory (which compiles), before you can proceed. We will guide you through this.

The next step is to create the BlackBerry WebWorks PhoneGap project.

```
$>D:
$>cd d:\PhoneGap-plugin\phonegap-1.1.0
$>ant create -Dproject.path=D:\PhoneGap-Plugin\BB-Plugin-Test
```

This will show you the directory shown in Figure 9–13.

Name	Date modified	Type	Size
lib	11/2/2011 1:49 AM	File folder	
www	11/2/2011 1:49 AM	File folder	
build.xml	11/2/2011 12:59 AM	XML Document	11 KB
project.properties	11/2/2011 1:02 AM	PROPERTIES File	2 KB

Figure 9–13. *PhoneGap BlackBerry project structure*

Now let's ensure whether our plug-in class compiles.

```
$>cd D:\PhoneGap-Plugin\BB-Plugin-Test
$>ant build
```

If the above step shows some compilation errors in HelloWorldPlugin, you need to

1. Fix those compilation errors.

2. Delete the project at D:\PhoneGap-Plugin\BB-Plugin-Test.

3. Recreate the project using Ant create -Dproject.path=D:\PhoneGap-Plugin\BB-Plugin-Test.

4. Check the compilation using "ant build."

Now that you have gone past this issue of compiling a blank Java plug-in class, let's put some code in it.

Following is the complete code for the plug-in class (this is quite similar to the Android plug-in). We expose an action name "hello," and we expect an argument named "name." Given that someone calls out the action "hello" with the name "Rohit," we return back "Hello Rohit! The time is <current time>."

```java
package com.phonegap.plugins;

import com.phonegap.api.Plugin;
import com.phonegap.api.PluginResult;


import java.util.Date;
import com.phonegap.json4j.JSONArray;

public class HelloWorldPlugin extends Plugin {

        private static final String ACTION_HELLO="hello";

    /**
     * Executes the requested action and returns a PluginResult.
     *
     * @param action      The action to execute.
     * @paramcallbackIdThe callback ID to be invoked upon action completion.
     * @paramargsJSONArry of arguments for the action.
     * @return            A PluginResult object with a status and message.
     */
public PluginResult execute(String action, JSONArray data, String callbackId) {
        PluginResult pluginResult=null;
    if (ACTION_HELLO.equals(action)) {

                String name;
                    try {
                            name = data.getString(0);
                            String result = "Hello " + name
                                    + "! The time is "
                                    + (new Date()).toString();
                            pluginResult =
                                    new PluginResult(PluginResult.Status.OK,
result);
                            returnpluginResult;
                    } catch (Exception e) {
                            pluginResult =
```

```
                                                            new
    PluginResult(PluginResult.Status.JSONEXCEPTION,
                                                "missing argument name");
                  }

            } else {
                        pluginResult =
                            new PluginResult(PluginResult.Status.INVALIDACTION,
                                            "Allowed actions is hello");
            }
        return pluginResult;
    }

    /**
     * Called when the plug-in is paused.
     */
    public void onPause() {

    }

    /**
     * Called when the plug-in is resumed.
     */
    public void onResume() {

    }

    /**
     * Called when the plug-in is destroyed.
     */
    public void onDestroy() {

    }

}
```

Note that you have to dump the modified HelloWorldPlugin.java again in PhoneGap's framework as shown in Figure 9–13. You also have to delete and recreate the project using Ant create -Dproject.path=D:\PhoneGap-Plugin\BB-Plugin-Test to test the plug-in.

Declaring the JavaScript Part of the Plug-In

Again, the JavaScript part of the plug-in is very similar to the Android JavaScript part of the plug-in. In this case, we declare everything in a function declaration and call it as well.

```
(function () {
    var HelloWorld = function () {
            return {
                hello: function (message, successCallback, errorCallback) {
                    PhoneGap.exec(successCallback, errorCallback, 'HelloWorldPlugin',
    'hello', [message]);
                }
            }
        };
```

```
PhoneGap.addConstructor(function () {
    // add the plug-in to window.plugins
    PhoneGap.addPlugin('simpleplugin', new HelloWorld());

    // register the plug-in on the native side
    phonegap.PluginManager.addPlugin('HelloWorldPlugin',
'com.phonegap.plugins.HelloWorldPlugin');
    });
})();
```

The first step is to create a JavaScript object named HelloWorld and declare a function in it by the name of "hello." This function internally calls a PhoneGap-registered service, which in turn calls the actual native class.

Now that we have this object, which will be called from our HTML, we need to register this object as a PhoneGap JavaScript plug-in. We also need to map the service name "helloworldplugin" to the class "com.phonegap.plugins.HelloWorldPlugin." All this is done inside the PhoneGap.addConstructor() call.

We use PhoneGap.addPlugin() to map the "simpleplugin" name to the JavaScript plug-in object. This exposes the plug-in as windows.plugins.simpleplugin.

Then we use phonegap.PluginManager.addPlugin() to map the Service name to the actual Java class.

This completes the part where we create our JavaScript part of the plug-in. We will put this JavaScript inside the project's www directory.

Calling the Plug-In

To call the plug-in, we modify the index.html file present in the project's www directory.

This is very similar to what we did earlier for Android and iPhone.

Following is the code snippet used to invoke our plug-in:

```
window.plugins.simpleplugin.hello(
    document.getElementById("name").value,
    //success callback
    function (message) {
        document.getElementById("output").innerHTML = message;
    },
    //failure callback
    function () {
        log("Call to plugin failed");
    }
);
```

As we did earlier, we supply the name; in this case, the name comes from an input type text element. Then we provide a success callback and a failure callback. In the success callback, we set the return value in a div with id "output."

Here is the complete code for the index.html page:

```html
<!DOCTYPE html PUBLIC "-//W3C//DTD HTML 4.01 Transitional//EN"
"http://www.w3.org/TR/html4/loose.dtd">
<html>

    <head>
        <meta http-equiv="Content-Type" content="text/html; charset=UTF-8">
        <meta name="viewport" id="viewport" content="initial-scale=1.0,user-
scalable=no">
        <script src="json2.js" type="text/javascript">
        </script>
        <script src="phonegap-1.1.0.min.js" type="text/javascript">
        </script>
        <script src="helloworld.js" type="text/javascript">
        </script>
        <script type="text/javascript">
        function log(message) {
            document.getElementById("log").innerHTML =
document.getElementById("log").innerHTML + "<br>" + message;
        }

        function onDeviceReady() {
        }

        function sayHello() {

            window.plugins.simpleplugin.hello(document.getElementById("name").value,
            //success callback


            function(message) {
                document.getElementById("output").innerHTML = message;
            },
            //failure callback


            function() {
                log("Call to plugin failed");
            });


        }

        // register PhoneGap event listeners when DOM content loaded


        function init() {
            document.addEventListener("deviceready", onDeviceReady, true);
        }



        </script>
    </head>

    <body onload="init()">
        <h1>
            Simple Plugin Demo
```

```
        </h1>
        <table border="1">
            <tr>
                <td>
                    Enter Name
                </td>
                <td>
                    <input type="text" name="name" id="name">
                    </input>
                </td>
            </tr>
            <tr>
                <td>
                    <b>
                        Output:
                    </b>
                </td>
                <td>
                    <div id="output">
                    </div>
                </td>
            </tr>
            <tr>
                <td colspan="2">
                    <button id="hello" onclick="sayHello();">
                        Say Hello
                    </button>
                </td>
            </tr>
        </table>
        <div id="log">
            ...
        </div>
    </body>
</html>
```

The last step is to run the WebWorks BlackBerry project. Go to command prompt, go to the project directory, and run the following command:

```
$>ant build load-simulator
```

This will open the BlackBerry simulator, and you can see our application running inside it. Enter a value in the text box and hit the button. You will see the result shown in Figure 9–14.

Figure 9–14. *PhoneGap plug-in output*

Sharing the BlackBerry PhoneGap Plug-In

To share the plug-in, you need to publish two files

1. helloworldplugin.java

2. helloworld.js

Add the above to the documentation on how to invoke the plug-in from JavaScript.

Conclusion

Although JavaScript is a fast and flexible language for developing a cross-mobile application, JavaScript has certain inherent limitations when implementing complex processing and background work. Sometimes it's necessary to use native code to perform the heavy lifting.

PhoneGap's architecture allows us to extend its plug-in to introduce the native code for our PhoneGap application.

Index